The Road Taken

By the Same Author

The House with Sixteen Handmade Doors: A Tale of Architectural Choice and Craftsmanship

To Forgive Design: Understanding Failure

An Engineer's Alphabet: Gleanings from the Softer Side of a Profession

The Essential Engineer: Why Science Alone Will Not Solve Our Global Problems

The Toothpick: Technology and Culture

Success Through Failure: The Paradox of Design

Pushing the Limits: New Adventures in Engineering

Small Things Considered: Why There Is No Perfect Design

Paperboy: Confessions of a Future Engineer

The Book on the Bookshelf

Remaking the World: Adventures in Engineering

Invention by Design: How Engineers Get from Thought to Thing

Engineers of Dreams: Great Bridge Builders and the Spanning of America

Design Paradigms: Case Histories of Error and Judgment in Engineering

The Evolution of Useful Things

The Pencil: A History of Design and Circumstance

Beyond Engineering: Essays and Other Attempts to Figure Without Equations

To Engineer Is Human: The Role of Failure in Successful Design

The Road Taken

THE HISTORY AND FUTURE OF AMERICA'S INFRASTRUCTURE

Henry Petroski

BLOOMSBURY

NEW YORK · LONDON · OXFORD · NEW DELHI · SYDNEY

Bloomsbury USA
An imprint of Bloomsbury Publishing Plc

1385 Broadway	50 Bedford Square
New York	London
NY 10018	WC1B 3DP
USA	UK

www.bloomsbury.com

ISBN: HB: 978-1-63286-360-7
        epub: 978-1-63286-361-4

Library of Congress Cataloging-in-Publication Data
Petroski, Henry.
The road taken: America's imperiled infrastructure / Henry Petroski.
New York: Bloomsbury, 2016. | Includes bibliographical Reg. references and index.
2015018674| ISBN 978-1-63286-360-7 (hardcover: alkaline paper) | ISBN
978-1-63286-361-4 (ePub)
Roads—Economic aspects—United States—History. | Roads—Government
policy—United States—History. | Roads—United States—Maintenance and
repair—History. | Roads—United States—History. | Interstate Highway
System—History. | Infrastructure (Economics)—United States—History. | United
States—Economic conditions—1945- |
TECHNOLOGY & ENGINEERING / History. | POLITICAL SCIENCE /
Public Policy / General.
HE355.3.E3 P48 2016
388.10973—dc23

2   4   6   8   10   9   7   5   3   1

Typeset by RefineCatch Limited, Bungay, Suffolk
Printed and bound in USA by Berryville Graphics Inc., Berryville, Virginia

To William Henry

Contents

Preface

TODAY THERE IS A heightened awareness of the importance of our physical infrastructure to the economic health of the nation and the well-being of its citizens, not to mention to our national pride. Bridge and building collapses, water main breaks, sinkholes, deteriorating roads, and more have propelled infrastructure to hot-topic status. Heated debate in Washington about how to finance the maintenance, restoration, and replacement of our aging infrastructure has also kept the subject in the news. However, tight economic times, partisan politics, and public skepticism about whether our public works are being honestly and properly built and cared for have made it difficult to find the resources to do what truly needs to be done. The infrastructure problem is sure to grow in importance and interest, but unless our nation's attitude towards it changes, it is not likely to be solved anytime soon.

This book is about our infrastructure—its nature, its history; its pluses, its minuses; its funding and its financing; and its future. The nature and history of anything should be part of objective reality, but what a particular author chooses to include or exclude in describing and telling a tale can obviously bias the story. I have tried to present the infrastructure and issues relating to it in an evenhanded way, but it will naturally be from one traveler's point of view, which means that my education and experience as an engineer and lifelong student of the quotidian will have a lot to do with what I know and see, what I hear and read, how I evaluate and interpret it, and what I emphasize and what I pass over.

In conventional treatments of our highway infrastructure, roads and bridges receive the bulk of the attention. This book contains its share of roads and bridges, but additionally, it considers in some detail the parts of roads that are often overlooked: their seams and their edges, their potholes and their fringes. The centerlines, curbs, and gutters are considered, as are guardrails and medians, road signs, and traffic signals. These finishing details of the hardware are complemented by the software represented by the conventions of driving and the rules of the road, which are also treated in this book. These intangible adjuncts to highway infrastructure are every bit as important as are the tangible grade, bank, and curve of the pavement in making it safe and efficient to ride upon. And how well the details are attended to—in their design, construction, and maintenance—can tell us a lot about how well the underlying road itself is made and maintained. As with everything else designed, the success or failure of a piece of infrastructure is ingrained in the details.

Whether we like it or not, infrastructure is steeped in politics and we vote on issues relating to it accordingly. We vote for roads and against potholes; for fixing our bridges and against the taxes to do so; for pork and earmarks but not necessarily for the common good. We vote by anecdote, by what we want and need and see and feel daily in our own neighborhoods, towns, cities, and regions. It is often said that all politics is local; when it comes to roads and bridges, it is also state and federal, for those governments are the principal sources of regulations and funds to plan, design, build, and maintain capital investments. It is where roads lead and what bridges span—and whether they are perceived to do so well or poorly—that feed the anecdotal gossip and influence the vote. At present, in one way or another, all roads seem to lead to and from Washington, although since the federal government has been finding it increasingly difficult to match revenue for and spending on infrastructure, it has been signaling that it leans toward throwing the problem into the hands of the states. Whether or not the states are prepared to shoulder the full responsibility for their infrastructure remains to be seen.

IN THIS BOOK, I have chosen chapter titles that allude to Robert Frost's "The Road Not Taken." This poem, which opened Frost's 1916 collection, *Mountain Interval*, was written at a time when the automobile was young and roads were far from what we know them capable of being today. However, like all brilliant literature, the poem is essentially about timeless thoughts and themes. The infrastructural dilemmas faced by citizens and their legislative representatives today are fundamentally little different from the choice that a fork in the road presented to Frost's traveler through the woods a century ago. It behooves us all to keep such humble individual experiences in mind as our nation contemplates seemingly impossibly complex choices.

Our nation and its states are at a fork in the road today, and which path they choose to take in dealing with our decaying infrastructure will indeed make all the difference in whether our roads, highways, and bridges—and the economy that so relies upon them—can truly be revitalized. At the time of this writing, we are at a tipping point, in that serious decisions are being made that will influence how we think about infrastructure and our funding of it. By presenting historical background on how our roads, bridges, and traffic habits have come to be as they are, this book describes how past choices—both good and bad— have put us in the position we now are with infrastructure, and also presents the background to and current status of fundamental issues involved in contemporary legislation and practice related to infrastructure. It is imperative that we take into account the successes and, especially, the failures of the past in choosing the path to the future. When our children and grandchildren look back years hence, they will know that our choices made all the difference for us and for them.

H.P.

August 2015

The Road Taken

GIVENS, CHOICES, FORKS, SIGHS

I GREW UP IN an urban environment that epitomized infrastructure old and new. Naturally, like any city, New York had streets and side-walks and houses and churches and factories and stores. But, being a financial and cultural center, it also had office towers, skyscrapers, museums, concert halls, and libraries—all on a large scale and in great numbers. And it had bays and harbors and rivers and creeks and canals. It had docks and piers and ferry terminals; bridges and overpasses and underpasses; tunnels and railroads and subways and elevated trains called els, which took us to beaches and boardwalks and amusement parks, as well as to Ebbets Field and Yankee Stadium and the Polo Grounds. The city was crisscrossed by highways, mostly called expressways and park-ways, but it also had alleys and mews that recalled the rows of stables of horse-and-wagon days. There were stations and terminals and airports and heliports. You name it, New York had it—and it still does, though not necessarily in the condition I knew as a child and young man.

Growing up in Brooklyn, I roamed the streets and avenues in my neighborhood of Park Slope, ranging as far up the hill as Prospect Park—a tranquil island full of hills and meadows in the middle of a sea of asphalt and concrete—whose green undergrowth my friends and I explored at our leisure, especially in the summer. To get to the park, each day I could zigzag along a different route through the rigid grid of streets and avenues lined with brownstone row houses and other mostly domestic structures that crowded the sidewalks. If the day was nice enough for a walk to the park, it was nice enough to air out the house and, probably, have laundry drying on the clotheslines strung from window to window and window to pole across the diminutive back-yards. The higher I got on the slope, the grander the brownstones became and the fancier were the curtains billowing out of the open windows and the laundry hung by help. Naturally, the epitome of houses and views stood facing the park, directly across from the low stone wall that delineated it.

On some blocks there were isolated architectural gems and oddities, solitaires set among the more familiar cheek-by-jowl building types. One of note was a Second Empire mansion that stood on Ninth Street near Fifth Avenue. It was built as a country retreat by William B. Cronyn, a wealthy Wall Street merchant, in the mid-1850s, when this section of Brooklyn was still rural and punctuated by the occasional farmhouse. That changed with the development of the farmland itself and, no doubt, anticipation of the completion of the Brooklyn Bridge, which would occur in 1883, thus establishing the first fixed link between Brooklyn and Lower Manhattan; infrastructure has that kind of effect on land and neighborhoods alike. In 1888 the Cronyn house was bought by Charles M. Higgins, the manufacturer of patented "American India Ink." It and he prospered, and a decade later Higgins added a five-story brick building in the Romanesque Revival style behind the house that the business had outgrown. When I knew it, the high-fenced factory compound reached completely through from Eighth to Ninth Street, and when the gates were open we kids took the factory yard as a short-cut to and from the Avon, then a family movie theater with Saturday shows for kids.

Infrastructure has a way of imprinting itself upon us, and the phys-ical surroundings in which we grow up often become blueprints against which we gauge the world as adults. My paradigm for walking up a hill—any hill—remains to this day the trek up Park Slope among the rows of brownstones broken up only now and then by a freestanding mansion or small factory among the homes. And the goal of the climb, idyllic Prospect Park atop the hill, remains the touchstone against which I pit all parks and places of recreation. As for the Higgins ink factory, as

a child I thought more about the consequences of being caught tres-
passing on its grounds than I did about what went on in the tall brick
building. As a high school student I found myself taking a drafting
course whose list of required supplies included Higgins India ink. But
only toward the end of the course did we learn how to use the ink that
came in the bottom-heavy bottle with the eyedropper cap. We dripped
the ink between the blades of our pens and compasses to trace over
pencil lines and so made the impressions of plans we had formed
permanent. Our classroom exercises followed the order in which real-
world infrastructure was conceived, planned, and drafted with a pencil
and then written into legal documents with waterproof ink.

As children, we also ventured down the slope from the park, until our
progress was checked by the stagnant and odorous Gowanus Canal. We
generally stayed upwind of the adjacent warehouse area, except in the
weeks leading up to the Fourth of July when we were shopping for fire-
works. We would enter the storefront of a wholesaler under the pretense
of picking up boxes of penny candy for my friend's father's store and
exit with half bricks of firecrackers in brown paper bags. There were
several schools and playgrounds within walking distance of the street
on which my friends and I lived, but we preferred the street itself and
nearby parking lots for our proving grounds. We also used the streets
for games. At first simply tag and hide-and-seek, but later pitching
pennies, flipping baseball cards, and playing marbles, catch, handball,
stoopball, and stickball. Traffic was light after school and during the
summers, and we were seldom interrupted by a car or truck coming
down the one-way street. I cannot remember ever seeing a pothole; if
there had been one, I am sure we would have designated it home plate or
second base, depending on its location.

Playing in the street brought us in proximity with still more infrastructure, including manhole covers and sewer grates that were
peepholes into the underground world of pipes and conduits that
carried everything from essential water to unwanted waste. The hidden
supply lines surfaced now and then in the form of fire hydrants, which—
although located on the sidewalk—restricted parking in the street in
front of them. They also provided obstacles to leapfrog over, could

stand in as first or third base, and gave dogs an object to defile. The fireplugs were always right beside the curb, that distinguishable but otherwise undistinguished piece of infrastructure between the gutter and the sidewalk. Wherever we walked or played, we were treading on, over, or along infrastructure. Where a subway line ran beneath the street, sidewalk grates allowed the tunnels to inhale and exhale with each passing train. Infrastructure was in the air; it was ubiquitous.

But if it was not something we used as an instrument or backdrop for play, we hardly noticed it. Only when some piece of infrastructure was disturbed by adults, such as a condemned building being demolished or a flooded street being dug up to get at some buried pipes, did we take note. And then it was mainly to watch the activity, to watch the men at work and not what they were working on. Putting up a new building was not so interesting, because the process went so much slower than tearing the old one down. But generally there was not much construction going on in my well-established neighborhood in Brooklyn, where a run of walk-up apartment buildings was interrupted only now and then by the odd curb cut leading into a factory yard or alleyway. There were few garages because there were few automobiles. Those families who did own a car either parked it on the street or in a rented space beneath the elevated subway tracks nearby.

When our family grew, we moved to the suburbs, which for us meant to the farthest reaches of Queens, New York's easternmost outer borough. In fact, we moved to within just a few blocks of the Cross Island Parkway, which separated the borough of Queens from Nassau County and the rest of Long Island. Our house was not far from Idlewild (now JFK) Airport, and we soon learned that under certain weather conditions one of the approach flight paths was right over our house. Nevertheless, when the weather was good enough for us to play outside, my friends and I played touch football on the wide grassy stretch beyond the shoulder of the parkway or, when there were not enough of us to make up teams, we just sat on the grass and watched the cars go by, vying to name their makes, models, and years, and getting really excited when we saw a brand new model for the first time. Mostly, though, the streets—among the most basic elements of urban and suburban infrastructure—were our playing fields, whether for football, stickball, track and field, or rodeo bicycle riding.

When I was older, I borrowed my father's car on Saturday evenings to drive to see friends on Staten Island, a good thirty miles away via the Belt Parkway (which, as its name suggests, runs around the shore of Queens and Brooklyn); the parkway took me to the Staten Island Ferry terminal. The Verrazano-Narrows Bridge was under construction at the time, and I endured the long drive and ferry ride in part to follow the bridge's progress, which went much slower than a building's. However, since I passed the construction site only every few weeks at most, to me the span was advancing in leaps and bounds. When it opened in 1964, this structure—the last of the great New York City bridges designed by the distinguished engineer Othmar Ammann and the last major public works project overseen by master builder and power broker Robert Moses—surpassed the Golden Gate Bridge as the longest-spanning suspension bridge in the world. Today, it remains the longest in the Americas.

My adventures with New York City infrastructure—especially its roads and bridges—continued as I commuted between my home in Queens and college in the Bronx, essentially traveling each day from one of the southeasternmost parts of the city to one of the northwesternmost, a distance of almost twenty-five miles via boulevards, parkways, expressways, and a toll bridge. The vehicular commute was made by carpool. As traffic slowed down on the approach to the toll plaza, the day's driver mock-ritualistically held out an invisible collection basket, expecting to feel a quarter drop into his palm for each of the three or four passengers he was carrying that day. Since the bridge toll itself was just twenty-five cents each way, the additional quarters were understood to be for gasoline, which then cost about thirty cents a gallon.

The bridge we crossed was the Bronx-Whitestone, which was designed also by Ammann and opened just in time to carry automobiles and their passengers to and from the 1939 World's Fair. The aesthetically sleek structure was flexible to the point of its roadway undulating in the wind when conditions were just right—or wrong. Several contemporaneous suspension bridges exhibited a similar flexibility, and because of this in November 1940 the Tacoma Narrows Bridge in Washington State collapsed into Puget Sound during a storm. By the time my fellow commuters and I were using the Bronx-Whitestone, it had been retrofitted with heavy steel trusses that all but obscured a great view of the Manhattan skyline. Even so, when we were caught in stopped traffic on the bridge, I could feel it still heaving up and down like a breathing animal. And if I sighted through the truss on a distant stationary object—like a power pole or a steeple—I could confirm the bridge's movement.

All the while we commuters were using the Bronx-Whitestone Bridge in late 1959 and throughout 1960, construction was proceeding on yet another Ammann suspension bridge, this one located barely two miles to the east, easily within rubbernecking distance. Since I was an engineering student, I paid close attention to every incremental advance in the steelwork, savoring the spectacle. To the untrained eye, nothing was happening; to the budding engineer, everything was. So it is with infrastructure. That which just sits in place and functions without incident calls no attention to itself; that which has unintended movement, malfunctions, collapses, or explodes catches the eye of everyone. To the average citizen, infrastructure is neither seen nor heard until it flashes or bangs and so makes the news.

When the Throgs Neck Bridge did open to traffic, our carpool altered its route to pass over the new span; enjoying the sweeping new approach to Queens was also a departure from routine. Within a couple of years of its opening, the bridge provided us with a direct connection to the new Clearview Expressway that roughly paralleled our old route but was less familiar and so less traveled. This was all happening in the years following the passage of federal legislation establishing the interstate highway system, but not all plans of that ambitious program were to be realized—at least in their original form. The Throgs Neck Bridge and Clearview Expressway interstate route ultimately was supposed to carry motorists all the way to JFK Airport, but the Clearview ended (as it still does) abruptly at an array of unremarkable traffic signals at Hillside Avenue in the middle of the neighborhood known as Queens

Village. We had to wend our way through local streets to complete our journey.

MY ENGINEERING DEGREE WAS my ticket to a broader and less crowded world. Graduate school in the Midwest introduced me to teachers, students, and scholars from Canada, England, Italy, Hungary, India, Australia, New Zealand, China, Japan, and elsewhere. I was naturally interested in their different cultures while at the same time marveling at the fact that we were all using the same mathematics and science, studying the same engineering and technology to design and build skyscrapers, bridges, automobiles, airplanes, and—increasingly at the time—rockets, satellites, and spacecraft. Although we often used different words for similar parts of the made world, we had the same things and concepts in mind. In the case of down-to-earth infrastructure, for example, we wished for well-designed, smoothly paved, pothole-free, and long-lasting roads that were clearly marked, congestion-free, and as safe as humanly possible.

My professional life as an engineer has given me the opportunity to travel around the country and around the world, where I have experienced infrastructure of enormous variety. Still, whether I have found myself circling a roundabout in Ireland or crossing a bridge between Denmark and Sweden, the principles of traffic safety and control are fundamentally so similar that it takes hardly any time to acclimate to idiosyncrasies of handedness or local practice. Some regional imperatives exist, such as looking right before crossing a road where cars drive on the left; but even if all transportation infrastructure is ultimately local, it is but part of a universal system of roads and their appurtenances that gird our planet.

And everywhere on earth there is a situation that arises at every crossroads and every fork in the road: Which way to proceed? The answer is usually straightforward: Forge ahead, follow the crowd, take the main way, do as has been done. But every now and then one of our ancestors has paused at the intersection and thought about choice. If he was an engineer (in fact or even just in spirit), he may have wondered whether the design of the intersection (and perhaps the choices it might engender) could be improved for more efficient and safer traffic flow. Indeed, he may have lingered and sketched in the dirt or on the back of an envelope an idea to be mulled over later that day or evening. Of course, even travelers who were as far from being engineers as they could possibly be have also paused at a fork in the road and wondered, What if I took the other way?

THE POET ROBERT FROST was certainly no engineer, but he did contemplate the consequences of choice—in infrastructure and in life. His poem "The Road Not Taken" is about a traveler who was passing through some woods one morning when he came upon a fork in the road. Realizing that he could not go both ways, he stopped to decide which to take. He looked as far as he could down one road, but he could not see beyond where it turned in the undergrowth. He chose to take the other road, not because he could see any farther down it, but because as far as he could tell it seemed to be more inviting. Even though in the vicinity of the fork the two roads appeared to be equally worn and covered with newly fallen yellow leaves, a ways down the second "was grassy and wanted wear."

The traveler decided to keep the first road for some future journey. But even as he was deciding, he knew that, because "way leads on to way," it was unlikely that he would ever return to the fork. He also knew at the moment he made his decision that someday he would look back upon it and reflect on how his taking the less-traveled road had made all the difference in his life. However, like so many powerful poems, Frost's ends with a measure of ambiguity. He tells this story "with a sigh," but is it a sigh of relief, tiredness, yearning, or sadness? After all, the title of the poem is "The Road Not Taken."

America is now at a fork in the road representing choices that must be made regarding the nation's infrastructure. This is not the first time we have had to pause to decide the way forward, of course. For streets and highways alone, we have had to deal with the dilemmas of concrete versus asphalt pavements, of white versus yellow road markings, of

cantilever versus suspension bridges, of quality versus quantity in building, of elevated highways versus underground tunnels for traffic, of
electric motors versus internal combustion engines for vehicles. The
list goes on and is ongoing. Contemporary dilemmas include spending
money on fixing potholes or building up speed bumps, levying taxes
based on volume of fuel used or miles driven, promoting public or
private investment in infrastructure.

The United States now has more than four million miles of roads and
bridges, much of which was built for an earlier time, is now in poor
repair, and continues to become more and more congested. Potholed
and traffic-jammed roads mean that it takes commuters longer to drive
to and from work; it takes truckers longer to deliver raw materials and
goods from mine to plant to supplier to factory to warehouse to store;
and it takes everyone longer to pay off repair bills for wheel alignment
and damaged suspension systems, not to mention time wasted and ire
raised in slow-moving or stalled traffic. According to the Texas A&M
Transportation Institute, traffic congestion alone costs the country $121
billion annually in lost time, which amounts to more than $800 per
driver. For truckers, the cost in wasted time and diesel fuel consumption
is about $27 billion annually. These are among the consequences of
neglected or inefficiently funded infrastructure.

And it is not only roads and bridges that present us with conundrums
and dilemmas. So many of America's airports, seaports, waterways,
dams, water supplies, mass transit systems, high-speed rail networks, and
other commonplaces of modern life are either old and inadequate or
verging on the nonexistent. Seaports up and down the East Coast have
had to be modified to handle the larger cargo ships that will come through
a widened Panama Canal. New York's La Guardia Airport, gateway to
the glamour of Wall Street, Broadway, and Fifth Avenue, has been
described as "dingy" and called "America's worst." No less a supporter
of our infrastructure than Vice President Joe Biden, who as a senator
famously commuted by Amtrak between his home in Wilmington,
Delaware, and Washington, D.C., has angrily declared, "It is just *not
acceptable* that the greatest nation in the world does not have—across the
board—the single most sophisticated infrastructure in the entire world."

The historical record is clear, in that choices have always had to be made, but it can also be as ambiguous as Frost's reflective traveler about what has been and is the better choice. When about a century ago there were horse-drawn wagons competing with horseless carriages for space on the streets of our already crowded cities, the streets were littered with wet, mushy, smelly, unsanitary equine urine and excrement. The vaporous exhaust of motor vehicles must have seemed at the time to be the answer to a prayer. It took little more than a half century to realize that we had only traded one form of pollution for another, something realized with a sigh long after the seemingly technologically advanced, apparently sanitary, and logically sound choice had been made.

In this book, the historical infrastructural choices that have been made—sometimes willy-nilly—with regard to the design and selection of roads and bridges, as well as the invention and development of paving materials, traffic-control devices, and even driving conventions, serve as case studies on how right or wrong decisions can be and how long it can take to realize which is which. It is from the failures that we can take lessons learned, from the successes that we derive hope for the future.

Everyone knows that choices have consequences, ranging from benign to disastrous. But the consequences are hard to foresee when there is so much undergrowth in the way. Political debates, especially when partisan, can be so cluttered that it is difficult to see the trees, let alone the forest. Reflecting on some of the seemingly less consequential or even inconsequential choices that have been made—whether stop signs should be red or yellow, whether the red or green light should be topmost in a traffic signal, whether a suburban street should be called a drive or a road—can help us think through larger choices: whether a deteriorating highway should be paved this season or next, whether a new bridge should be of a familiar utilitarian design or a so-called signature one like the Brooklyn and Golden Gate bridges, whether an old bridge should be demolished or repurposed. Road and traffic commissioners in large cities and small towns alike are constantly facing such choices.

Choice pervades solutions to infrastructure problems, as it does those to virtually all engineering, economic, political, and aesthetic ones. By understanding how historical choices have been made—under what circumstances, under what technical and financial constraints, and under what hidden and overt political pressures—we can better understand what is involved in making key choices that we are faced with today. Regardless of the ultimate implications of a choice, understanding the process by which it was arrived at can be invaluable to individuals and groups faced with analogous choices.

At the present time and for the foreseeable future, increasingly difficult decisions about infrastructure are being made and will be faced by local, area, and state government officials and governing bodies. Numerous states will be faced with and tempted by proposals to enter into partnerships with private investors that will provide much-needed revenue to balance budgets, but at the risk of abdicating control of public property to private interests, which may or may not be wise. Knowledge of the case studies discussed in this book may help prevent a political misstep.

At the federal level, representatives in Congress will be facing possibly game-changing choices, among which are whether to continue the Highway Trust Fund—and how to replenish it—or to turn responsibility for paying for roads and bridges over to the individual states. Not only must the long-term result be considered here, but also the manner in which states are to be weaned over time from the federal money that they have counted on for so long. If the federal government were to cease taxing gasoline at the pump, for example, how should the states respond? Should they just replace the federal tax with an increased state fuel tax large enough to fund their roads and bridges? If a state tax were to be levied, should it be on the volume of fuel purchased—as it is now—or on the price of the fuel? Or should vehicles be taxed on the basis of miles driven? The choice made can make the difference between balancing a state budget or not.

Throughout this book, examples of how a wide variety of roads and bridges have been planned, designed, financed, built, maintained, and managed provide both inspiring stories and cautionary tales of what has and what has not worked when it comes to street and highway infrastructure. Every town, municipality, region, and state has its own idiosyncrasies, of course, and so the case histories are not necessarily directly applicable to a similar situation in a different time and place. However, by looking beyond the specifics to the common features of the process, a great deal of wisdom can be derived from what has been tried, and how and why certain choices have been made. It behooves us to learn from the process, if not the product.

The choices regarding infrastructure that we in the voting booth and our government representatives in the legislative chambers make today will affect the quality of our lives. Everyone involved in the decision making, whether as a citizen voter or as a congressperson, bears responsibility for making informed and correct choices. None of us should wish to foresee having a grandchild or great-grandchild on our knee ages and ages hence and having to tell him or her about our choices with a sigh that is at all ambiguous. We should all take a deep breath now and confront our infrastructure problems head-on. And this can best be done by understanding the history and present reality of the technology, economics, and politics of building and maintaining especially our roads and bridges, arguably the most visible, directly encountered, and representative parts of our infrastructure. Armed with informed knowledge of the problem and with the power of the ballot, citizens must support their government representatives in finding the courage to act decisively to align America's infrastructure with its aspirations.

The Road Not Taken

WORDS, REPORTS, AND GRADES

"INFRASTRUCTURE" IS A RELATIVELY new word for an absolutely ancient concept. Today, infrastructure connotes the sum of a society's physical improvements and denotes the public works (that is, structures and systems like roads, bridges, and water supplies that serve the public) as well as the works of private enterprise (for example, the fiber-optic, wireless, cellular, and other information and communication networks) that enable a civilization to function in a civilized way. "Infrastructure" is appropriately a Latinate word, given that the Romans were infrastructural geniuses. Their triumphs, like the Appian Way and the Pont du Gard, are still respected and admired after two millennia. But "infrastructure" as a common English word has a surprisingly recent etymology, with the present meaning being traced only from 1927, according to my desk dictionary. Google's graph of the word's appearance over time shows it to have been virtually unused until the early 1950s, after which usage grew exponentially until about 2000, when it reached a plateau. Ironically, the word has matured at the same time that much of what it designates is nearing the end of its useful life.

This explicit word for the concept of the underlying foundation or basic framework of any system or organization is believed to have had its origins in the French army. The forced interaction between English and French speakers during World War I led to the intercultural transfer of the concept to the British military. Military usage, in which "infrastructure" connotes the bases and camps needed to maintain, support, and deploy troops, was further reinforced in the late 1940s and early 1950s following the development of the North Atlantic Treaty Organization, and this also introduced the word into American English. Subsequently, urban planners adapted it to civilian use.

"Infrastructure" remained an obscure word even into the late 1980s, with its relative newness highlighted when it was placed between quotation marks in a *Wall Street Journal* article. The *Los Angeles Times*, which years earlier had also put the word in quotes, even called the term a "fancy label." However, the *New York Times* seemed to be more comfortable with the word, printing it without quotation marks as early as 1981 in an opinion piece lamenting the deterioration of means of production in U.S. private industry. The word was treated similarly in a 1988 op-ed essay by then Arkansas governor Bill Clinton, who echoed growing concerns about the nation's public facilities: "Our infrastructure is just barely adequate to support our current level of economic activity, and our current rate of infrastructure improvement and investment falls vastly short of tomorrow's needs." Decades later, the word still seems foreign to some television news people, who pronounce it "infa-structure," as if the prefix did not contain the letter *r*.

The term "public infrastructure" was used sparingly in a 1981 report from the Council of State Planning Agencies, perhaps more to avoid constant repetition of the term "public works" than to imply anything more. The report, titled *America in Ruins: Beyond the Public Works Pork Barrel*, was written by Pat Choate and Susan Walter. An economist, Choate had already had a wide range of policy and management experience with state and federal governments; years later he would become Ross Perot's Reform Party vice presidential running mate in the 1996 U.S. presidential campaign. Walter was associate director of the council, which was affiliated with the National Governors Association, which clearly had vested interests in public works and how they were cared for and funded. At the time, public works were considered by more than a few citizens to be synonymous with "pork barrel" spending, for the report emphasized that "well-conceived public works are not 'pork barrel.'" Had Choate and Walter written their report a couple of decades later, they might have employed the term "earmark."

America in Ruins was a primer on the nature, condition, and importance to the economy of infrastructure by any name. With supporting data presented in numerous charts and tables, Choate and Walter made the case that "public works play a crucial role in the creation of national wealth and productivity growth." They also analyzed the interplay between capital investment in infrastructure and cycles of economic recession. The growing practice had been to cut back on infrastructure spending as a means of balancing budgets in lean times, whereas Choate and Walter argued that just the opposite should be the practice to stabilize budgets. They wrote of the condition of the infrastructure that "the greater part of the decline reflects the growing habit of government at all levels to cut back on construction, rehabilitation and maintenance in order to balance budgets, hold down the rate of tax growth and finance a growing menu of social services."

Road construction has always had an enormous ripple effect on the economy. In 1922 alone, within a period described as "a golden age of road building," more than ten thousand miles of highways were constructed with federal aid. According to Earl Swift, writing in *The Big Roads: The Untold Story of the Engineers, Visionaries, and Trailblazers Who Created the American Superhighways,*

> Roads ballooned into a huge employer, providing jobs not only for those actually building them, who numbered in the hundreds of thousands, but also for an army of men who made road-laying gear and provided the raw materials. More than 200 American companies made cement, 127 made paving brick, and 42, asphalt; another 380 provided crushed stone, and 340 shipped sand and gravel. The various public officials involved in roads numbered eighty thousand.

In *America in Ruins* Choate and Walter recognized that cost-cutting practices of the 1970s were "undermining efforts to revitalize the economy and threatening, in hundreds of communities, the continuation of such basic services as fire protection, public transportation, and water supplies." They warned that public facilities in America were wearing out faster than they were being replaced. They also emphasized the potential of public works spending to carry the economy through periods of recession such as the one that the country was then experiencing. Steady spending on infrastructure maintenance was likened to a "flywheel effect," in which fiscal energy was stored to supplement a sagging economy. At first the report received little attention, but as the recession worsened, Choate and Walter's alert got front-page coverage in the *New York Times*, along with cover stories in *Time* and *Newsweek*.

The report was reissued two years later with a different subtitle, *The Decaying Infrastructure*. This removed the perhaps off-putting idea of pork barrel from the book's cover and at the same time raised the word "infrastructure" and the condition of the nation's essential physical plant to the level of conscious use that persists in the United States today. Choate and Walter touched a nerve with their indictment of government for its declining investment in public works. But, the book emphasized, it was difficult to know the condition of the national infrastructure and to estimate the level of funding needed for maintaining and improving it.

WE TEND TO BE oblivious to much of our infrastructure, even when it is in plain sight, until something goes wrong with it. The power lines that dip from pole to pole beside the street in front of my house were never more visible to me than after they were severed during an ice storm and lay powerless on the ground for a week. Now, years after I watched the linemen restring the system, I see the cables almost every time I look out my front window and I realize how vulnerable they are. And I see the pole between my house and the one next door to be as out of plumb as it was on the day our power was restored. Had service never been interrupted, I might never have noticed it.

A century ago, in our larger cities, wires and cables of all kinds dipped low over streets crowded with horses, wagons, carriages, omnibuses, trolleys, automobiles, trucks, and pedestrians. Looking down from an upper-story window was like looking at the street through a sagging horizontal screen formed of electric, telegraph, telephone, and

This photograph of a New York City street was taken in the wake of the blizzard of 1888. The countless number of wires hanging overhead, accented by the snow accumulated on them, were not only a sign of the technology of the times but also a blight on the cityscape. Telephone and other wires would, of course, eventually be buried in larger American cities. However, in much fewer numbers they remain overhead in smaller towns and older suburban neighborhoods.

clothes lines that were strung from pole to pole, from pole to building, from building to building, from building to pole—so densely packed as to darken the street scene beneath them with their shadows.

In time, of course, many of those lines were buried under city streets, joining water mains, sanitary sewers, gas lines, steam pipes, and pneumatic tubes. When it came time to construct a subway tunnel, the task might have appeared akin to pushing a paper straw through a bowl of spaghetti. Today, things are a bit more organized underground, but workers operating a backhoe to dig a trench to lay fiber-optic or some other new kind of cable not infrequently cut through some prior infra-structural installation.

Buried infrastructure in many cities can be of the order of a century old. Cast-iron water mains burst spontaneously, causing the resulting craters to fill and overflow. In the United States, each day on average 650 water mains fail and seven billion gallons of water are lost through leaks, resulting in about 16 percent of our clean drinking water being wasted. In New York City, some parts of the water supply system are said to have been operating at full capacity since their installation and so have never been emptied for inspection. A clogged condensation trap in a steam line buried beneath Lexington Avenue in Manhattan was blamed for the 2007 explosion that left one person dead and a dozen or so injured. The steam pipe itself, which had been installed in 1924, was not found to be weakened by age; but its traps, which were designed to bleed off unwanted condensation, were fouled with debris from gasket sealant used in the pipe joints.

Infrastructure can be counterproductive and, needless to say, danger-ous when not maintained or functioning properly. In 2010 in San Bruno, California, the spontaneous explosion of a natural gas pipeline killed eight people, injured many more, and destroyed three dozen homes. A National Transportation Safety Board investigation found the cause to be defective welds in steel pipe that had been installed in 1956, before sophisticated X-ray inspection techniques were available for quality control. When the pressure in the 30-inch-diameter pipeline was increased to serve a growing demand, the already flawed pipe was further weakened. It was just a matter of time before it blew. Almost five years after the event, Pacific Gas and Electric, the owner of the pipeline, was fined $1.6 billion for violating safety standards. Approximately half of the fine was to pay for improvements in the utility's pipeline system.

In the wake of the publication of *America in Ruins*, the state of the nation's infrastructure understandably became a topic of further studies, including a three-year one conducted by the congressionally chartered National Council on Public Works Improvement. This five-person council's landmark final report, titled *Fragile Foundations: A Report on*

America's Public Works, which was issued in 1988 to the president and Congress, did not contain the word "infrastructure" in its title, but the word did appear in the opening paragraph of the report's summary of findings, which echoed much of *America in Ruins*:

> The quality of a nation's infrastructure is a critical index of its economic vitality. Reliable transportation, clean water, and safe disposal of wastes are basic elements of civilized society and a productive economy. Their absence or failure introduces an intolerable dimension of risk and hardship to everyday life, and a major obstacle to growth and competitiveness.

The council, going a bit further than Governor Clinton had, stated that, on balance, "our infrastructure is inadequate to sustain a stable and growing economy. As a nation, we need to renew our commitment to the future and make significant investments of our own to add to those of past generations." Among the features of *Fragile Foundations* was a "Report Card on the Nation's Public Works," in which each of eight categories of infrastructure (highways, mass transit, aviation, water resources, water supply, wastewater, solid waste, and hazardous waste) was assigned a letter grade based upon performance and capacity.

This provided a condition assessment of the kind called for by Choate and Walter. Highways received the best grade, a C+, reflecting the fact that "improved pavement conditions" had resulted from increased state and federal gasoline taxes and from increased spending on operations and maintenance. Such factors as increasing congestion, inadequate road expansion in areas of growing population, and the aging condition of roadways and bridges kept the grade from being better. The report also noted that the Highway Trust Fund, which had been established in 1956 to provide a source of money for maintaining roads like the interstate highway system, had a "sizeable cash balance," implying that much more money was available to be spent to improve the nation's highways. Indeed, in 1987 the unspent balances in trust funds for highway, aviation, transit, and waterways (subsequent legislation had expanded the concept) stood at almost $24 billion, while the infrastructure (valued at almost $1 trillion at the time) that the funds were intended to help maintain continued to deteriorate. Investment in existing public works had taken a backseat to new construction and environmental concerns.

In addition to assigning letter grades to the various categories of public works, the report made qualitative and quantitative assessments of the overall infrastructure and its needs. Most major categories were judged to be "performing at only passable levels." The council called for the government, private industry, and the public to make a national commitment "to vastly improve America's infrastructure." The kind of commitment the council had in mind was expected to require a doubling of the annual national investment by government in new and old public works, which in 1985 stood at about $45 billion. But at the same time, the report cautioned that the problems with the nation's infrastructure "cannot and should not be solved through a crash program." A "sustained effort" was needed.

For a report that concerned itself exclusively with infrastructure and its future, *Fragile Foundations* contained some curiosities. For one, before the next-to-last page of the summary volume of the three-volume report, there was virtually no mention of engineers or engineering. Only on that penultimate page was it noted that "enrollments in schools of civil engineering—traditionally the mainstay of the public works profession—are declining significantly just as many senior engineers are reaching retirement age." What was true in the 1980s remains a concern more than a generation later. Indeed, the report's conclusion that "the nation has a shortage of technically competent personnel to meet future requirements of the public works profession" remains essentially valid, as does the report's recommendation that "government agencies and private organizations should conduct an aggressive and sustained campaign to encourage young people to enter the profession."

A DECADE LATER, IN 1998, the American Society of Civil Engineers (ASCE) issued its own "Report Card for America's Infrastructure," which expanded and refined some of the categories in the National

Council on Public Works Improvement report and renamed others. The highways category was split up into roads and bridges, the water resources category emphasized dams, and the new category of schools was added. While the ASCE cautioned against a direct comparison of its grades and those issued in the earlier report card, the letter grade conceit begged for one. As can be seen in the summary table located at the end of this chapter, the infrastructure received poorer grades in all but two categories: solid waste received the same C– grade, and the grade for mass transit improved from C– to C. The release of the first ASCE report card attracted considerable media coverage nationally and regionally and the institution of the infrastructure report card became solidly established.

Overall, assuming the grading systems of the National Council on Public Works Improvement and the American Society of Civil Engineers were comparable, the nation's infrastructure went from an average grade of C to one of D over the course of ten years. Perhaps the ASCE's initial grading of the infrastructure was a bit harsh. The society issued a new report card three years later, in 2001, and on it grades improved or stayed the same in all but two categories: mass transit and wastewater. Two new categories were added: navigable waterways and energy, both of which received the grade of D+. The power generation capacity of the nation was not felt to be keeping pace with demand, and the transmission infrastructure relied on old technology. The report card also raised the question of long-term reliability. Overall, the 2001 report card gave America's infrastructure a D+, and it was estimated that a five-year investment of $1.3 trillion was needed to improve it.

A progress report issued by the ASCE in 2003 did not assign new grades but indicated trends. No category received an upward arrow signifying improvement, and only five of the twelve categories were holding steady. As was the case two years earlier, the evaluations were based on "condition and performance, capacity vs. need, and funding vs. need." Congested roads had been estimated to be costing the nation's economy $67.5 billion annually in wasted fuel and worker productivity. Of the nation's 600,000 bridges, more than a quarter were considered to be structurally deficient (that is, closed or kept open only

with speed and weight restrictions because of deterioration) or functionally obsolete (having narrower lanes or shoulders than called for by
current standards). The dissemination of such trends and numbers was
effective in calling attention to the unsatisfactory state of the nation's
infrastructure.

Within two years, a wholly new report card was issued by the ASCE.
Three new categories were added in the 2005 edition: public parks and
recreation, rail, and security. The first two each received the grade of
C−, the last an "Incomplete." Seven of the dozen established categories
received lower grades than they did on 2001, and the overall infrastructure grade was assessed as a solid D. Indeed, except for bridges maintaining a grade of C and solid waste one of C+, none of the carryover
categories was graded higher than a D+. In the meantime, the price tag
for bringing the infrastructure up to acceptable condition had risen to
$1.6 trillion over a five-year period.

The next report card was scheduled to be released in March 2009, but
the ASCE feared that would be too late to influence the stimulus
bill that was being considered in the earliest days of the new Congress
and the new administration in Washington. The new president's
call for including significant infrastructure spending in the legislation
under consideration prompted the ASCE to release the grades on its
2009 report card about two months early. The grades were announced
at a press conference held the week after the presidential inauguration.
The news delivered was not good. Only one category—energy, as
manifested in the national power grid—had shown improvement since
the 2005 report card. Most others just maintained their mediocre
to poor grades, with three (aviation, roads, and transit) actually
dropping a notch. The new estimate of the cost for repairs was
$2.2 trillion.

A new category, levees, received a grade of D−. That this category
was not included on report cards before the devastation that Hurricane
Katrina had recently wreaked on New Orleans demonstrated how
complacent even engineers and professional societies can be about
infrastructure. It took that tragedy to call attention to the importance
and sometimes fragile condition of the nation's 100,000 miles of levees,

over 85 percent of which were privately owned and maintained with unknown reliability. A good number of these dated from a half century earlier and were constructed to protect farmland from flooding. In the meantime, however, residential developments had grown up behind these levees, and so they were now relied upon to protect people as well as crops. According to the report card, they were doing it poorly.

One category that does not appear in report cards is buildings. These, especially large office buildings, are certainly important components of a nation's infrastructure and its preparedness to serve commercial ends. Next to bridges, perhaps, today's world-class super-tall buildings, which often serve multiple purposes, tend to be the most visible aspects of the infrastructure of large cities and contribute significantly to the public perception of urban infrastructure. In addition, buildings that house institutions like museums, performing arts centers, and monuments are an integral part of the cultural infrastructure by which a nation is known. While many such buildings are public works in the sense that they are conceived, designed, and built with federal, state, and local government funds, they are supported to a considerable extent also by private philanthropy.

The category of security, which had been introduced in the 2005 evaluation—the first one issued after the terrorist attack on New York's Twin Towers—was dropped from the 2009 report card because engineers had "begun to look at security in the context of infrastructure's overall resilience—or the ability to withstand and recover from both natural and man-made hazards." As a consequence, resilience was taken into account in grading each category.

The ASCE evaluations, analyses, and indictments of public policy regarding the nation's infrastructure that had been going on for three decades were by now familiar-sounding refrains. They were echoed in President Obama's speeches leading up to his inauguration and to the passage of the stimulus package represented by the American Recovery and Reinvestment Act of 2009 written and passed in the first weeks of the new administration. Still, the money that was allocated to construction-related projects of all kinds—estimated to be as low as $30 billion and as high as $130 billion, about a quarter of which was to go to highway projects—was considered by industry observers to be woefully inadequate to the task. Even accounting for the extraordinary stimulus money, infrastructure spending in 2009 was expected to drop by more than 4 percent from the previous year. Infrastructure remained merely a fancy label for the stuff driven over, buried, and otherwise ignored compared to more glamorous aspects of the economy, society, and civilization that it makes possible. Of the $41 billion in federal funding for transportation infrastructure authorized in 2013, only 6 percent went for new roads and bridges, with the balance shoring up aging and deteriorating ones.

In spite of a virtually constant refrain about the dismal condition of our aging infrastructure, in the ASCE's 2013 report card no category received a grade lower than it did in 2009, and the overall grade improved from D to D+. Still, the estimated investment required to bring all infrastructure up to acceptable levels was put at $3.6 trillion by 2020.

A new category, ports, was added in 2013, no doubt reflecting the fact that when the expansion of the Panama Canal was complete, larger container ships would bring a new volume of commerce to the East Coast, but only if its commercial ports in cities like Miami, Savannah, Charleston, Norfolk, Baltimore, New York, and Boston could accommodate them not only in water depth but also in dock and crane capacity. Longer and wider post-Panamax ships would also be higher, requiring some unusual infrastructure adjustments, such as raising by sixty-four feet the roadway of the Bayonne Bridge, located between New Jersey and Staten Island and in its original configuration limiting access to Newark Bay. The unprecedented engineering feat of replacing the roadway of a steel-arch bridge was estimated to cost $1.3 billion. And not only larger cargo ships drive infrastructure modifications: the 190-foot clearance beneath the twenty-seven-year-old Sunshine Skyway Bridge will not allow passage of the towering cruise ships of tomorrow, which could cost Tampa, Florida, dearly in tourist business. Constructing a new port in outer Tampa Bay was estimated to cost as much as $650 million. Building a new, higher bridge and demolishing the old could cost as much as $2 billion, and the project would take six years to complete.

In anticipation of larger container ships passing through an enlarged Panama Canal, many modifications in the infrastructure of East Coast ports became necessary. The Bayonne Bridge, shown here as it was originally built, was the longest steel arch bridge in the world when it opened in 1931. It was an unprecedented engineering challenge to raise its roadway to allow taller ships to gain access to docks located in Newark Bay.

The quadrennial report cards issued by the ASCE, which encourages its sections to issue report cards for their own states or regions, certainly do bring attention to our infrastructure needs and show them to be broad and complicated. The categories of roads and bridges, perhaps most tangible and relatively easy to grasp, serve as surrogates for the ensemble. However, even the casual observer, seeing these and virtually all infrastructure categories receiving the mediocre and poor grades in the C and D range since the late 1990s, cannot help but wonder why our nation has not redoubled its efforts to improve our roads and bridges and by implication the rest of our infrastructure over that period. It is as if collectively we are like a poor student, incapable of or resistant to

REPORT CARDS FOR AMERICA'S INFRASTRUCTURE

	NCPWI 1988	ASCE 1998	ASCE 2001	ASCE 2003	ASCE 2005	ASCE 2009	ASCE 2013
Highways	C+						
Roads		D−	D+	↓	D	D−	D
Bridges		C−	C	↔	C	C	C+
(Mass) Transit	C−	C	C−	↓	D+	D	D
Aviation	B−	C−	D	↔	D+	D	D
Water resources	B						
Dams		D	D	↓	D	D	D
Water supply	B−						
Drinking water		D	D	↓	D−	D−	D
Wastewater	C	D+	D	↓	D−	D−	D
Solid waste	C−	C−	C+	↔	C+	C+	B−
Hazardous waste	D	D−	D+	↔	D	D	D
Schools		F	D−	↔	D	D	D
(Navigable) Inland waterways			D+	↓	D−	D−	D−
Energy (national power grid)			D+	↓	D	D+	D+
Public parks & recreation					C−	C−	C−
Rail					C−	C−	C+
Security					I		
Levees						D−	D−
Ports							C
Average grade	C	D	D+		D	D	D+
Investment Need (in trillions of dollars over five years)	0.2	1.3	1.6		1.6	2.2	3.6 by 2020

ASCE = American Society of Civil Engineers
NCPWI = National Council on Public Works Improvement
Grade definitions: A = Exceptional, B = Good, C = Mediocre, D = Poor, F = Failing, I = Incomplete
Progress report trend definitions: ↔ = No change, ↓ = Worsened condition

learning the lessons that have been so clearly laid out to us, here in the form of deteriorating highway pavements and bridge decks. As adults, we complain about them, we demand of our politicians that they do something about them, but we stop short of voting for the funding that we are told it will take to do so. Of course, we are reluctant to do so at least in part also because of past experience that does not appear on the well-publicized series of report cards. Do our elected officials and bureaucrats spend our money wisely? Do they make the right decisions not for the short term of the election cycle but for the long term that roads and bridges can reasonably be expected to last without having to be repaved or replaced in the interim? These are the kinds of questions that report cards alone do not answer.

3

Roads

DIRT, STONE, WOOD, CONCRETE, ASPHALT

As LONG AS ANIMALS and people have walked the earth, they have worn paths through fields and forests. Wild buffalo may have carved out the first North American trails, defining migration routes from the grazing areas of one season to those of another; Native Americans followed the bison trails to move to and from their own seasonal hunting grounds. These hoof and foot paths would have gotten muddy in the rainy season and slippery when covered in snow and ice, and such primitive thoroughfares throughout the world would have called out for improvement. Ancient civilizations leading up to empires like Rome's found it in their bureaucratic and military interests to develop harder, wider, and more technologically sophisticated roads. The Roman army was certainly more efficiently deployed over stone-paved routes such as the Appian Way.

The Romans began the construction of a road by defining its course and width with parallel furrows, then digging down about three feet to excavate, as much as possible, a stable trench. Into this were placed first a course or two of large flat stones set in lime mortar. Onto this approximately fifteen-inch-thick base were placed small stone fragments also embedded in mortar that filled the voids. Next, a concrete consisting of pieces of broken pottery and brick, crushed stone and gravel, all mixed into more lime mortar, was compacted to a thickness of about six inches. On top of that, a final layer of irregular but closely fit stones of about six-inch thickness was placed. The Romans called this hard top

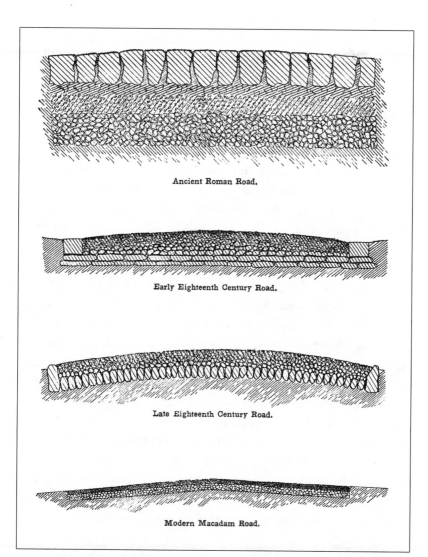

Ancient Roman Road.

Early Eighteenth Century Road.

Late Eighteenth Century Road.

Modern Macadam Road.

Early roads were very time- and labor-intensive to build, but like any technology they and the methods of making them evolved. The Industrial Revolution brought an increased use of new materials, methods, and machines. Today, traffic on interstate highways generally can continue to proceed even as the roads are being resurfaced.

surface the *pavimentum*, from which the English word "pavement" derives. Portions of such ancient roads have endured for more than two thousand years. To reproduce such a road in modern times would be prohibitively expensive, and it would have drainage problems.

The construction of stone-paved roads was obviously very labor and materials intensive, and the Romans and less expansive civilizations looked for simpler and cheaper ways to put a dry face on dirt and mud paths. One solution was to lay whole logs across the width of the roadway and fill the spaces between the rounded sides of the logs with sand and clay, thereby achieving somewhat of a smooth surface but one that did not last, as the logs began to shift and the sand filtered down through the interstices. Nevertheless, such quickly constructed roads continued to be used into modern times, especially over low, swampy land, where they were a great improvement over dirt and mud, even though they were bumpy to ride on and hazardous for horses to walk upon. These so-called corduroy roads were used during the Civil War by General Sherman, whose men advanced a dozen miles a day in their march through the Carolinas, and during World War II by the Germans and Soviets on the Eastern Front. However, such roads remained serviceable for substantial periods of time as long as the logs were buried in peat or other soil that slowed the rotting in the wood. Split logs laid flat side up might have provided a smoother alternative, but splitting the logs evenly and maintaining them in a stable position would not have been easy. The plank road, which the Romans also used, worked better, in that the flat-faced boards generally stayed put where they were laid down. But exposed logs and lumber were subject to deterioration and rot, and so alternatives to the vulnerable material were sought. Engineers have constantly learned how to make roads that can be built more economically and perform more effectively, but even the best of them today may not last as long as surviving Roman paved roads.

The use of river gravel or broken stone could produce a fairly good road surface, especially after it had been compacted, but such pavements were easily damaged by sharp objects like walking sticks or heavy concentrated loads like those imposed by narrow-wheeled wagons. Horses pushing off with their hooves, especially during their faster gaits, would also have dislodged loose gravel and stone chips, thereby pitting the road surface and opening it up to accelerated deterioration. When it rained, the part of the road that was not undermined or washed away would have its holes and ruts fill with water, making it difficult if not impossible for a traveler to judge how deep they were. Such roads were clearly far from ideal and required regular maintenance.

Roads may be visually distinguished by their surface, but it is what is beneath the surface that is at least as important to their proper functioning. Modern road design had its origins in France, whose land transportation system in the eighteenth century was considered the world's finest. The individual most responsible for this was Pierre-Marie-Jérôme Trésaguet, who came from a family of engineers. The construction of a Trésaguet road began with the digging of a trench down to solid subsoil, which was shaped to contain a slightly raised center ridge known as a crown. Next, large flat stones were placed vertically against the sides of the trench and smaller flat stones fitted tightly between them to form an arch across the road base. Still smaller stones were pounded into any voids in the arch. Atop this was spread a layer of walnut-sized gravel, and atop it a layer of slightly smaller but sharp-edged crushed stone, which when compacted provided a solid all-weather pavement for horses and wagons.

Scottish engineers led the development of roads that improved upon Trésaguet's design. Foremost among them was Thomas Telford, who made so many lasting contributions to the British infrastructure in the late eighteenth and early nineteenth centuries—in the form of major roads, bridges, canals, and harbors which remain critical parts of the nation's transportation network—that his friend the poet Robert Southey gave him the nickname "Colossus of Roads." Telford improved on the French road by using higher-quality stone and by using angular broken stone to fill voids between larger stones. He also used a mixture of gravel and crushed stone to form the pavement surface. Finally, whenever possible, he raised the pavement above the adjacent ground surface to drain water away from the road.

Crushed stone typically required hand labor, men and boys hammering on rocks and stones to break them into highly faceted pieces that

locked together when compacted into a road surface. Quality control
was provided by an inspector with a ring gauge: If a piece of broken
stone did not fit through the iron ring, it was rejected and broken into
smaller pieces. In the absence of a mechanical gauge, a piece of stone
could be tested by seeing if it could be passed through rounded lips into
the mouth. Telford's method was improved upon by John Loudon
McAdam, who had spent a dozen or so years as a businessman in New
York during the American Revolution before returning to Scotland in
1783 a rich and influential individual. He took an interest in roads after
becoming involved with the Ayrshire Turnpike Trust. His prior experience in running a colliery that supplied the British Tar Company made
him familiar with the sticky substance. Like Telford, he advocated
raising roads above the prevailing grade by layering rocks and gravel to
provide proper drainage. However, unlike Telford, McAdam did away
with the expensive arch-like foundation. Atop a base of large stones he
placed crushed stone and gravel bound together with tar. He gave the
road a camber so that water would run off to the sides and be carried
away in drainage ditches, thereby leaving the road foundation dry. This
construction method came to be referred to as macadamisation, which
was shortened to macadam.

Early nineteenth-century roads in Britain developed a reputation for
being superior. In the early 1820s, the Danish scientist Hans Christian
Ørsted embarked on a tour of European countries to present and debate
his astounding experimental results that showed an electric current
flowing through a wire to have an effect on a magnetic needle held
nearby, thereby establishing a connection of some kind between electri-
city and magnetism. The achievement earned him election to the Royal
Society in London, membership in which he naturally accepted. After
crossing the English Channel, Ørsted could not help but notice how
pleasant he found the coach ride to London. Indeed, he wrote that the
turnpike roads over which he traveled were "as smooth as a dance floor."

In America, the first road to be macadamised was also the first to be
federally funded. The National Road, constructed between the 1810s
and the mid-1830s, provided a route to western settlements. It was
known as "the Main Street of America" and it eventually became a part

of U.S. 40. By the end of the nineteenth century, about half of all American paved city streets were done either in gravel or macadam, which was essentially a technically improved way of using gravel, and main roads in Europe were also built according to McAdam's system.

McAdam's road-making technique did leave small gaps between the tar-coated crushed stone and gravel, however, and this shortcoming challenged other engineers to seek ways to make the surface smoother. One engineer, Richard Edgeworth, solved the problem while at the same time finding a use for the dust that was a byproduct of breaking stones with hammers and otherwise crushing them into small angular pieces. Edgeworth mixed the stone dust with water and filled the gaps in the pavement with the paste, making for a much smoother surface. The process came to be known as "water-bound macadam." Another way of smoothing the surface was to coat it with tar, resulting in "tar macadam," or "tarmac."

WHEN RAILROADS BEGAN TO connect cities, their locomotives were soon banned from the central streets. The loud smoke-belching engines were seen not only as noisy and dirty but also as boiler explosion and fire risks. As a result, a locomotive was uncoupled from the train of cars at what was considered a safe distance from city center and horses were hooked up to pull the cars the rest of the way. Thus, city pavements were interlaid with rails, which came to be used also for horsecars, an early form of intracity mass transit. Since the rails were the most level and smooth part of the street, teamsters had their horses pull wagons so that their wheels rode the rails as much as possible and thereby gained a smoother and faster ride.

Even where railroads or horsecars did not run, it was desirable to provide a smooth surface for wagon wheels to ride upon. This idea led to the development of what were variously known as stone rails or wheel tracks. In the early 1830s, when stagecoach and turnpike companies feared that the new steam-powered railroad would capture their trade, the road between Albany and Schenectady, New York, for example, was improved with a form of stone pavement that had long

been used in cities in Italy, England, and Scotland. Rather than make
the whole width of a road smooth, these cities had laid down two
ribbons of wide, flat stone bedded in gravel and separated center to
center by the distance that wagon wheels on an axle were apart, leaving
the rest of the road roughly paved. The stones were approximately
eighteen- to twenty-four-inches wide and four inches thick. Both top
and bottom faces were finished flat, so that when the top became
worn, the stone could be turned over in place to provide a new, smooth
surface. By keeping the wagon wheels on the stone pathways, a swift
and comfortable speed could be maintained. When stone rails on the
Albany-to-Schenectady road that had been laid down in 1834 were
photographed in 1901, they showed the wear of almost three-quarters
of a century, with grooves that resembled water channels. However, the
stones were clearly still serviceable.

*Since wagon wheels—especially narrow ones—cut into dirt and gravel
roads and led to the development of deep ruts, heavily traveled roads
were sometimes partially paved with lines of flat stones spaced
approximately the distance that the wheels were apart on a wagon
axle. Over time, iron-rimmed wheels wore ruts in the stone, which
then had to be turned over or replaced. This road in Ulster County,
New York, was built in 1862; the photo of it was taken in 1902.*

Using tar or asphalt to pave over a gravel road not only provided a smooth riding surface but also kept down the dust and protected the underlayment. But early asphalt roads were not durable and so had to be resurfaced regularly. In urban areas where near-constant traffic was very hard on streets, paving blocks were often used to produce a more lasting surface. The blocks could be made of stone—in which case they were termed "setts"—or wood. As can be experienced in historic areas of older cities today, streets paved with cobblestones or time-rounded setts make for a bumpy and noisy ride in a wheeled vehicle, but the nature of the pavement also slows the traffic down and hence makes for a longer-lasting road surface. Blocks made of wood set on the end grain or of granite result in a smoother and therefore quieter surface as long as they are well seated in sand or fine gravel so that any settlement is uniform.

Brick came to be a popular alternative paving material in the late nineteenth century, and by 1902 it covered about 1,300 miles of paved streets in the United States. At the time, brick was competitive with asphalt and produced a quieter (but not silent) pavement than cobbles or stone blocks. It also had a distinct advantage over wood blocks, which absorbed horse urine and so were odious and unsanitary. However, brick also had its problems, which included the noise factor in otherwise quiet neighborhoods, a relatively slippery surface, and a lack of durability and quality control. According to William Pierson Judson, writing in his treatise on city roads and pavements, "irregularly and imperfectly burned brick" that was "laid by incompetent contractors, under inexperienced city officials" failed needlessly and gave brick a bad name. Judson emphasized that paving bricks were entirely different from building bricks, starting from different materials and being made by different processes. He wrote further that calling the pavers "bricks" was misleading, for they were in fact tile. Properly made paving bricks do not absorb much water and so are unaffected by frost and "if formed of the best material properly treated will be tough, to withstand the blows of horses' toe-calks; hard to resist the abrasion of wheels, and strong to carry heavy loads." But not all bricks were made of the best material or properly fired, and so the whole category of paving bricks had to put up with the bad reputation earned by the few bad examples.

Judson was optimistic in 1902, however, believing that if for a new paving job a city engineer would simply choose bricks known to have performed well in other cities, a process of "natural selection" would drive out of the market the poorer grades of brick. Unfortunately, he did not seem to remember "Gresham's law," which states that bad money drives out good.

Regardless of what formed the surface of roadwork done at the turn of the twentieth century, concrete was typically used as a base of four- to six-inch thickness over which a two-inch layer of sand was evenly spread. It was onto this sand bed that granite, wood, brick, and other pavers were set, usually in a tessellating pattern. Where asphalt was to be the pavement surface, the concrete base was first covered with a binder comprising stone and asphalt steamrollered to a depth of about an inch and a half. The asphalt wearing surface was laid on top of the binder to a depth of a couple of inches.

The Romans used brick in building walls and concrete in creating vaulted ceilings. Concrete technology appears to have been lost between Roman times and the Industrial Revolution, but it was rediscovered in the middle of the nineteenth century when it was used to make rounded forms like flower urns and boats. By the end of the century modern concrete—a composite material consisting principally of cement, water, and sand, gravel, or stone—was being used for larger structures like buildings and bridges. The Alvord Lake Bridge, which opened in 1889 in San Francisco's Golden Gate Park, is believed to be the first reinforced-concrete bridge built in America. The modest twenty-nine-foot-span structure was designed by Ernest L. Ransome, an engineer and architect who patented a method of reinforcing concrete with twisted square or ribbed round steel bars. He had also promoted the construction of re-inforced concrete sidewalks, but left the City by the Bay in disappointment when it did not embrace his new technology more widely. He was vindicated when his park bridge, unlike the city's wooden structures, survived the 1906 earthquake in fine shape. The first American street to be surfaced in concrete is said to be Court Street in the northern Ohio town of Bellefontaine, done in 1891. In 1909, Woodward Avenue in Detroit became America's first mile-long stretch of concrete road.

Asphalt concrete—commonly called simply asphalt or blacktop—contains sand and stone bound by asphalt, the naturally occurring tarry substance that was known for centuries by its Latin names *asphaltum* and *bitumen* and that had long been used as an adhesive and sealant. The modern use of it as a road-paving material was conceived by a Swiss engineer by the name of Merian. He noticed that when carts loaded with rock bitumen bounced over bumpy roads, they lost some of the smaller pieces of their cargo. These fell into depressions and ruts in the road, and when compacted by traffic made for a better road. In 1849, Merian converted the chance occurrence into a controlled process and used it to pave a street in Switzerland. Soon streets in Berlin, London, and Paris were paved with asphalt. Among the earliest attempts in America occurred in New York City, when in 1869 a layer of asphalt was applied over granite blocks on Fifth Avenue. Unfortunately, black dust from the procedure soiled houses along one of the city's most prominent streets, and so the paving technique did not prevail.

A more successful application occurred the following year when a Belgian-born chemist turned engineer named Edward J. de Smedt paved the street in front of City Hall in Newark, New Jersey. Departing from the European practice, De Smedt employed material from an asphalt lake on the Caribbean island of Trinidad and heated it before mixing it with sand. A few years later, in preparation for the celebration of the U.S. Centennial, the De Smedt method was used by the Army Corps of Engineers to pave a stretch of Pennsylvania Avenue in Washington, D.C., and subsequently seventy miles of D.C. streets were also surfaced with the new paving material. By the end of the century, 1,500 miles of streets in more than one hundred American cities were covered with asphalt, which teamsters preferred to granite because its crevice-free surface offered less resistance to wagon wheels rolling over it, and the relatively soft pavement was gentler on horses' legs. Furthermore, whereas the grooves between granite pavers had to be scraped clean by hand, sanitation workers could employ more mechanical means to sweep a smoothly paved asphalt street. This advantage gained for asphalt the strong support of public health advocates. Bicyclists, who were among the most vocal early advocates for better roads and became strong supporters of the Good Roads Movement that originated in the late 1870s and flourished into the 1920s under early automobile enthusiasts, understandably also very much preferred asphalt over stone and gravel roads.

Sanitary and municipal engineers also favored asphalt at least in part because of its ability to allow for more efficient cleaning and quieter traffic flow. But although local governments had the right to pave and maintain streets, property owners abutting a road had traditionally been able to call for its paving and choose the kind of pavement they wished. Especially in crowded cities, the streets outside the downtown area served not only to carry traffic but also as gathering places and playgrounds when there were no nearby parks. As a result, the people living along a neighborhood street often wished to keep it unpaved or paved with gravel or cobblestones in order not to encourage traffic. However, the rapid rise of the automobile made it increasingly difficult to escape traffic completely. Fast-moving rubber-tired motor vehicles so tore up gravel pavements that they soon ceased to be a viable option in an urban area. Cobble and other stone pavements were too rough for the speed at which autos traveled, and so asphalt—which was backed by a wide range of influential groups, from teamsters to engineers—soon became the pavement of choice. This was especially the case when asphalt lakes and other naturally occurring sources of the stuff became readily available. These lakes, also known as tar pits, were formed where crude oil seeped to the surface and, after its more volatile components had vaporized, left behind a large puddle of thick natural asphalt.

Oil had been discovered in Pennsylvania in 1859, but asphalt made from it contained paraffin, which froze in the winter and led to the cracking and deterioration of roads paved with it. The petroleum from California and Texas oil fields that opened up in the early twentieth century did not have so large a waxy component and so was more suitable to making good asphalt artificially. However, experience working with the still relatively new paving material had not kept up with the increasing diversity of its sources and composition. As was the case with many technologies of the time, standards were almost nonexistent and so results were uneven. A breakthrough came from the inventor Frederick

J. Warren, who with his six brothers founded the Boston-based Warren Brothers Company, which exploited the use of a range of types and sizes of crushed stone in its asphalt mix.

Warren patented his pavement design in 1901 and used the brand name Bitulithic for it. The angular stone chips—collectively known as aggregate—interlocked more effectively in the patented pavement and so provided a considerable improvement on contemporary paving systems. In fact, the Warrenite-Bitulithic material was so successful that competitors sought ways to get around the patent, which called for aggregate graded in size from one-half inch to three inches. The circumventers did so by using aggregate smaller than one-half inch in size. These new pavements—whether patented or not—proved to be even more effective than gravel and macadam, which the pneumatic tires and higher speeds associated with automobiles had been causing to deteriorate at a rapid pace. The Warren brothers also developed specialized machinery to install asphalt pavements, and successive innovations resulted in automated machines that paved roads quickly and with a minimum of backbreaking shoveling by road crews.

Whatever the methods, materials, and machines used, by 1922 road building was declared by the editors of *Engineering News-Record* to be "one of the nation's great industries." Still, the question of concrete or asphalt was hotly debated through the 1920s and 1930s. In the meantime, toll highways were on the drawing board, and the choice between the two materials could make the difference between a successful and unsuccessful enterprise. The Pennsylvania Turnpike, a state project that opened in 1940 as the first American paved toll highway, chose concrete. The state of Maine took a different route, creating the Maine Turnpike Authority, an independent agency with responsibility for the toll road's design and construction. After a cost-benefit analysis, the authority chose to pave the road, which opened in 1947, with asphalt, a decision of which highway engineers across the country took note.

By the early twenty-first century, asphalt was in place on about 94 percent of the more than two million miles of paved roads in the United States, only 20 percent of which are under the jurisdiction of the individual states. The remaining 80 percent are maintained by counties, cities, and towns. One of the reasons that asphalt predominates today is that many of the concrete roads that were laid down as long as a century ago were, beginning in the 1980s, resurfaced with asphalt, ostensibly as a cost-saving measure. And, generally speaking, wherever asphalt made inroads, it continued to be used as a paving, patching, and repaving material.

4
–

Diverged

TRANSCONTINENTAL JOURNEYS AND INTERSTATE
DREAMS

IN 1919, DWIGHT DAVID Eisenhower, then a young lieutenant colonel in the U.S. Army, rode in a military convoy that traveled from Washington, D.C., to San Francisco. It was the service's first transcontinental journey by motor vehicle, and the purpose of the exercise was to promote the Army's Motor Transport Corps and demonstrate its defense mobility. Almost three hundred enlisted men and officers traveled by motorcycle, car, and truck in a caravan that at times stretched for three miles. Vehicles were draped with bunting of red, white, and blue, and were accompanied by a band sponsored by the Goodyear Tire & Rubber Company, which no doubt saw an opportunity to promote land travel and transportation in peace as well as war.

The convoy of eighty-one vehicles, which was seen off from Washington on July 7 with speeches from the secretary of war and numerous senators, took sixty-two days to cover the 3,250 miles of the route that carried it through Maryland, Pennsylvania, Ohio, Indiana, Illinois, Iowa, Nebraska, Wyoming, Utah, Nevada, and, finally, California. There were naturally crowds of well-wishers along the way, and there were speeches and festivities befitting the arrival of such a parade of military men and equipment in the towns it passed through. Perhaps not to be outdone by Goodyear, the tire magnate Harvey Firestone welcomed the traveling army to his estate in Akron, Ohio, where "a covey of young ladies dressed in gay frocks treated the men with a magnificent feast."

Gaiety and politicking may have slowed its progress somewhat, but the convoy nevertheless did set what was for the time and circumstances the remarkable pace of fifty-two miles per day, or about five miles per hour—not much faster than a brisk walk. Although excruciatingly slow by today's standards, it was about the best that the caravan could do at the time. The journey had to be undertaken in the heat of summer in order to ensure that the western mountains were passable. The season, however, also meant that what roads there were would be muddy after rain. Vehicles would overheat and otherwise break down. There were accidents with which to deal. Roads, where they existed, were often so rough that no significant speed could be achieved. Even the better roads were liable to be interrupted by a creek or river, hopefully one that was

In 1919 a U.S. Army convoy traveling across the country encountered numerous challenges, ranging from muddy country roads to flimsy bridges across creek beds and gullies. The sixty-two-day trip made a deep impression on Lieutenant Colonel Dwight D. Eisenhower, who would go on to make a lasting impact on America's infrastructure by promoting the establishment of the interstate highway system.

fordable or served by a ferry. If there was a bridge, it was often too light
for military vehicles to use without breaking it. (A supply officer with
the convoy said of old wooden bridges that "it wouldn't be a bad idea to
run over them and break them down" to demonstrate their weakness
and force their replacement.) Covered bridges were often too narrow or
had too low a clearance to allow military vehicles to pass.

It did not take a great leap of the imagination for Lieutenant Colonel
Eisenhower to conclude that the United States would benefit greatly
from an improved system of highways and bridges. Eisenhower's feelings on the matter of American roads would intensify during World
War II, when he would witness firsthand the great advantages of the
German Autobahn. All these experiences would inform Eisenhower's
infrastructure policy making when he became president. In the meantime, other champions of better roads across America would step
forward.

Although it may have been the first encumbered by heavy equipment
and hoopla, the Army convoy that so negatively impressed Eisenhower
was far from the first transcontinental driving experience. In 1903,
Horatio Nelson Jackson, a thirty-one-year-old Vermont physician and
novice driver, made a fifty-dollar bet that an automobile could traverse
the country. Jackson engaged Sewall K. Crocker, a mechanic and chauf-
feur, to be his traveling companion and backup driver. In a Winton
motor carriage powered by a two-cylinder, twenty-horsepower engine,
the two men set out from San Francisco and took sixty-three days to
reach New York via a northerly route, much of which was not paved.
The first woman to drive across the country was Alice Huyler Ramsey,
a twenty-two-year-old New Jersey housewife who—accompanied by
three other women—drove the 3,600 miles (only about 150 of which
were paved) from New York to San Francisco in only fifty-nine days in
1909.

The idea for a transcontinental road was put forward in 1913 by Carl
Graham Fisher, a bicycle racer and builder of a motor speedway on
the outskirts of Indianapolis. Fisher argued at the time that America's
highways were "built chiefly of politics," when the "proper material
is crushed rock or concrete." He proposed a "coast-to-coast rock

highway" that would run from Times Square in New York to Golden
Gate Park in San Francisco, and he hoped that it might be completed
in time for the opening of the Panama-Pacific International Exposition
in 1915.

Fisher proposed that the rock highway be funded by "automobile
barons" like Henry Ford, but Ford opposed the idea, arguing that if
private industry began to pay for good roads, the government would
never be expected to do so. Henry Bourne Joy, president of the Packard
Motor Car Company and one of the supporters of Fisher's idea,
suggested that the road be named the Lincoln Highway and be
dedicated to the revered president. In this way, the federal government
might be persuaded to divert to the highway project the $1.7 million
intended for a marble memorial to the president that was to be erected
in Washington. The transfer did not occur, but with the Federal Aid
Road Act of 1916, funds began to be made available to the states on a
matching basis.

According to the 1917 edition of the *Wanamaker Diary*—a combination daily planner, book of advertisements, and informative essays—
thirteen transcontinental highway "through routes" were then being
planned. About evenly divided between east–west and north–south
routes, they were only "in the first stages of permanent improvement."
In addition to the Lincoln Highway, east-west routes included the
Pike's Peak Ocean to Ocean Highway (New York to California),
National Old Trails (Washington, D.C., area to Los Angeles), Trail
to the Sunset and Santa Fe Trail (New York to San Diego), and Old
Spanish Trail (St. Augustine, Florida, to San Diego). North-south routes
included the Atlantic Highway (Calais, Maine, to Miami), Pacific
Highway (Vancouver to San Diego), Dixie Highway (Chicago to
Miami), and Jefferson Highway (New Orleans to Minneapolis-St. Paul).
A "diagonal automobile road across the country" was also in the planning stages. It was to be the longest one on the continent and was to be
given the pedestrian name of Savannah to Seattle Highway. At the time
such roads were being proposed, there were 3.5 million automobiles and
a quarter million trucks waiting to use them, but these numbers were
dwarfed by a national population of 21 million horses.

By 1925, there were so many named roads throughout the country that the named-highway system was becoming increasingly cumbersome to master and confusing to use. Individual states developed segments of these routes to maximize travel within their boundaries, thereby increasing the local and regional benefits of tourist dollars. Utah, for example, rather than completing a section of the Lincoln Highway, spent its highway money on the Arrowhead Route to Los Angeles. Also, it developed the Wendover Road as an alternative to the Lincoln Highway for travelers on their way to San Francisco. Even intrastate routes were difficult to comprehend outside one's familiar area. By 1917, the New York State Department of Highways had introduced a code for identifying main routes throughout the state. Bands of color on telegraph and telephone poles and other stationary objects, such as bridges, distinguished classes of routes: red indicated a principal east-west route, blue a north-south, and yellow a diagonal one. In so marking its roads, New York joined Massachusetts, Connecticut, Rhode Island, and New Hampshire in the practice, thereby making regional driving and touring more easily accomplished. At the same time, Wisconsin began replacing named trail and highway signs—which often consisted of little more than the color band painted on a pole—with numbered route markers of a distinctive shape. According to the August 1919 issue of the *Highway Magazine*, "Signs are the voices of the highway. They tell the traveler where he should go; how fast he may travel, and when he must beware of danger. Without signs the road is silent, unknown and menacing."

In 1925, the American Association of State Highway Officials, which dated from 1914, began to rationalize the system of roads and road signs across the nation. Henceforth, major routes through large stretches of the country were to be known primarily by numbers. The scheme introduced by AASHO (now known as AASHTO and pronounced "ash-toe" because since 1973 transportation officials have been included in the rubric) designated major east-west routes by numbers that are multiples of ten, with U.S. 10 being the most northerly route and U.S. 90 the most southerly. Major north-south routes, of which there were expected to be many more than east-west, were to have numbers ending in a 1 or a 5. Thus it was the location and alignment of U.S. 1 that gave it that designation—and not, as is commonly stated, that it was the first national highway. By AASHO's ordering principle, U.S. 1 ran along the East Coast between Florida and Maine; U.S. 101 was the coastal highway between California and Washington State. Under this scheme, the Lincoln Highway lost its identity as a single route, different sections of it being given different numbered designations; but devotees installed thousands of roadside markers identifying it as the highway dedicated to Abraham Lincoln.

President Franklin Delano Roosevelt had a direct and personal interest in the condition of America's roads. He was an ardent automobile driver, and when paralysis in his legs restricted his ability to drive, he devised a system of levers by which he could control the pedals with his hands and thereby enjoy driving himself around the countryside near Warm Springs, Georgia, where he went for hydrotherapy. Roosevelt's love of the road was exhibited in a 1936 memorandum in which he restated a route of his own devising connecting Shenandoah National Park, in Virginia, to Worcester, Massachusetts. The following year he called the chief of the Bureau of Public Roads to the White House to show him a map of the United States on which Roosevelt had inscribed three north-south and three east-west lines that he envisioned as a system of transcontinental toll routes.

Committees were appointed to look into the feasibility of an interregional highway system, which eventually led to a 1939 report that proposed a national highway system. The report, *Toll Roads and Free Roads*, was given special cachet when issued by the U.S. Government Printing Office as a "Message from the President of the United States." Among other things, the report concluded that, with the exception of a small number of highly traveled routes—in the Northeast, for example—toll roads were not feasible because they would never pay for themselves. The report also put forth a master plan for the development of a toll-free national "system of direct interregional highways designed to facilitate the long and expeditious movements that may be necessary in the national defense, and similarly wide-ranging travel of motorists in their own vehicles," which would eventually become realized as the interstate highway system.

Thomas Harris MacDonald had served as chief of the roads bureau from 1919 to 1939— becoming known within the highway community as "the Chief"—and then as commissioner until 1953. During the 1920s he had justified rural highway construction as being essential to "get the farmers out of the mud" that was synonymous with the roads that connected their farms to markets. In the late 1930s MacDonald, described as an "engineer's engineer" and "order personified," condemned the politically charged term "interregional" to the Washington dust heap when he declared it to have "absolutely no significance." But the long legislative battle over a national system of superhighways did not turn simply on a word. Much more important to the players involved were the relative roles of the federal and the state governments. A term like "interregional" was too blunt an instrument for such a battle, and conflicts between urban and rural preferences stymied legislation. The interregional idea may not have succeeded at the time, but it would be revised with gusto into an interstate system in good time. MacDonald had prepared a firm foundation for its eventual implementation.

Dwight Eisenhower assumed the presidency in 1953, bringing to the office his memories of difficulties with early cross-country driving. His appointees soon put forth proposals to develop in earnest a true interstate highway system. One advocated forming a National Highway Authority that would abolish state highway departments. Among the obstacles to realizing such a scheme was the fact that the federal government had no authority under the U.S. Constitution to build or maintain any roads at all—but it could pay for them. Ideas for financing a system of interstate roads ranged from all costs being borne by the states to complete federal funding. Eisenhower himself, being a fiscal conservative, at one time favored a system of toll roads that would be self-liquidating—that is, the tolls would last only as long as it took to pay off the debt incurred to construct them. In the end, compromises of all kinds had to be made.

Meanwhile, motor vehicles continued to multiply. In the mid-1950s, the president of the National Automobile Dealers Association could declare, "We have not built as many miles of highway since World War II as we have built miles of passenger cars." When the association met

in Washington, D.C., for an annual convention, the first lady, Mamie
Eisenhower, hosted a White House reception for car dealers' wives and
"was dazzled by the furs and diamonds the women wore," attesting to
the prosperity of the industry. More than 60 million motor vehicles were
registered in America at the time, and new roads were sorely needed.

WHAT WE NOW KNOW as the interstate highway system did not emerge
fully formed during the Eisenhower administration, a system of national
roads having been the dream of traffic engineers and presidents alike
for decades. Much of the planning, engineering, and legislative groundwork had been done in the 1930s and 1940s; what remained was the
development of a financial model that would satisfy both the federal
and state governments. After a couple of years of intense debate, the
Federal-Aid Highway Act of 1956 was finally passed, thus providing
the basis for the year 2006 having been called the fiftieth anniversary
of the Interstate Highway System. Among the enticements that
finally won over sufficient support for the legislation was the promised
miles of interstate highways around cities, where voters and hence their
representatives' votes in Congress were concentrated. President
Eisenhower signed the routes-for-votes bill without ceremony: he was
in Walter Reed Army Medical Center at the time, recovering from
surgery for ileitis. Since he was looking forward to being nominated
for a second term, he did not want to appear in public in a weakened
state. In the bill that Eisenhower signed, almost thirty-seven years to
the day since he had set out from Washington with the Army convoy
headed for San Francisco, $25 billion was authorized over twelve years
for constructing a National System of Interstate and Defense Highways.
It increased the federal gas tax to feed a Highway Trust Fund, put the
federal share of interstate construction costs at 90 percent, expedited
the process of acquiring rights-of-way, established standards for such
physical features as lane and median width, and set a completion date
of 1972.

Massachusetts Route 128 around Boston, with its limited access and
remote location from the congested city center, served as a model for

much of the interstate system. When it was first opened in 1951, Route 128 was referred to as "the road to nowhere"; but by the time serious planning for the interstates began, the economic development that Route 128 had attracted, especially around its exits, was undeniable. The promise of the interstate system was not fully realized by the legislative deadline of 1972, however, for opposition arose from groups ranging from environmentalists to neighborhood preservationists. The first fully completed transcontinental interstate highway was I-80, running from New York to San Francisco; its last section to be built, through Salt Lake City, was finally opened in 1986; but by then, since much of the 46,000 miles of interstate roadway had been built to 1972 standards and traffic expectations, it was already obsolete and in need of constant widening and upgrading, as interstate travelers have been all too well aware.

From the beginning of the interstate system, it was clear that an essential feature in making the network of highways user-friendly was a clear and distinctive system of numbering and signage that was readily readable and understandable to a driver moving at high speed. The numbering system adopted mirrored that of the U.S. highways. Major east-west interstate routes were given two-digit numbers that are multiples of 10, but with the lowest numbered (Interstate 10) being the most southerly and the highest (I-90) the most northerly. This scheme thus minimized the occurrence of nearby U.S. and interstate system routes carrying the same number. Major north-south interstate routes end in a 1 or 5, but with the lowest number (I-5) running along the West Coast and the highest (I-95) along the East. Lesser interstates were given even or odd numbers according to whether their predominant direction was east–west or north–south, respectively.

Triple digits designate interstate spurs or beltways that branch off from or separate from and reconnect to the main interstate routes. Thus, I-495 is the Capital Beltway around Washington, D.C., that allows I-95 through traffic to avoid the congestion in the central city. There is also an I-495 beltway around Boston, but there is little chance of confusion between identically numbered routes four hundred miles apart. Baltimore, however, is only about forty miles from Washington, so designating its beltway I-495 could have led to confusion. Thus, Baltimore's is I-695. The spurs I-295 and I-395 both lead into the heart of Washington from I-95 and its associated beltway. There are also interstate spurs numbered I-595 and I-895 in the Washington-Baltimore area, the former connecting the Washington beltway with Annapolis and the latter providing an alternate tunnel route under Baltimore harbor.

Establishing a rational numbering system solved only part of the problem of designating highway routes. To be truly useful, the numbers needed to be effectively displayed on distinctive signs, with the interstates clearly distinguished from the U.S. routes, especially in the Midwest, where there was a good likelihood of nearby ones bearing the same number. Route number signs for U.S. highways have since 1926 carried black numbers on a "U.S. shield" with a white background. After the interstate highway act was signed into law, the state highway departments were invited to propose designs for interstate route signs. Most of the states submitted concepts that incorporated the name of the state through which the interstate highway would pass, and the now-familiar red, white, and blue "federal shield" design represents a combination of the ideas submitted by Missouri and Texas, employing the top half of one design and the bottom of the other.

When in the mid-1950s standards for interstate highway exit signs were being developed, it was agreed that they should be readable at eight hundred feet, but the color scheme became a subject of some debate. White printing on a green background was recommended by an AASHO committee working in conjunction with the Bureau of Public Roads, but the newly selected federal highway administrator, Bertram Tallamy—who was actually color-blind—insisted that a dark-blue background, as used on the New York State Thruway with which he had been affiliated, was a superior choice. The committee representing the states was equally insistent that green was better. In the end, the issue was resolved by constructing a roadway section along which were erected differently colored signs directing drivers to "Metropolis" and "Utopia." Motorists who drove at 65 miles per hour past the signs to the make-believe destinations were polled on background color preferences.

At 58 percent, green was the clear choice over blue (27 percent) and black (15 percent). Now, of course, the familiar white-on-green signs are emblematic of the interstates and other major highways.

The readability of highway signs has long been of interest to transportation engineers, and the ubiquitous typeface so familiar to drivers for decades was dictated in the Federal Highway Administration Standard Alphabets for Traffic Control Devices. However, as the population was aging, it became increasingly obvious that there was going to be a growing percentage of drivers over sixty-five years old on the roads and that the vision of these drivers could be expected to decline as they aged further. Among the problems known to afflict such a population of motorists was difficulty in distinguishing some letters, especially under nighttime driving conditions. An early proposal was to increase the font size used on highway signs by about 20 percent. A larger type size would naturally require a larger sign area, which would have meant not only more costly signs but also more "visual clutter" on the roadscape.

In the late 1980s, Donald Meeker, a sign designer originally hired by Oregon for a project, teamed up with James Montalbano, a type designer, to improve the legibility of highway signs. They ended up modernizing and rationalizing the alphabet that the Federal Highway Administration (FHWA) used, which dated from mid-century and which they claimed "had never really been tested." The pair of designers focused on such details as redrawing the vowels *a*, *e*, and *u*, which "tend to 'fill in' under bright lights, making them indistinguishable," and also the vowel *i*, which can sometimes look like an *l*, as users of e-mail know all too well. The visibility of signs employing the new typeface of Meeker and Montalbano was compared with that of signs using the old one by erecting examples of each on a test track. The redesigned letters remained sharp from distances as much as 50 percent farther. The new typeface was christened Clearview; in 2004 it was granted interim approval by the FHWA, and examples of the signs were erected on sections of I-80 and I-380 in Pennsylvania. The improved legibility was estimated to reduce reaction time by as much as two seconds, which can give a significant advantage to a driver traveling at interstate speeds. However, the typeface was not added to the FHWA's 2009 edition of its manual of standards, apparently because its numerals and other typographical details had not been fully tested. In the spring of 2014 the agency indicated that it was looking to rescind its interim approval of the typeface. Infrastructural things can move slowly and indecisively even where interstate highways are concerned.

5

Yellow

CENTERLINES, STOP SIGNS, TRAFFIC LIGHTS

ONE SUMMER EVENING MY wife and I drove from Arrowsic, Maine, to Portland for dinner. We knew that thunderstorms were forecast, but summer storms are usually fleeting in the area and we saw no reason for them to keep us from celebrating our anniversary at a favorite restaurant. However, no sooner did we leave U.S. Route 1 in Brunswick and merge into the traffic on Interstate 295 than the rain began, first in big but sparse drops, then soon in waves of windblown sheets that came so swiftly and thickly that the wipers could barely sweep away one onslaught before the next arrived. At times, traffic slowed to 10 or 15 miles per hour as drivers struggled to keep the vehicle ahead of them in view but at a distance. Had it not been for the large red taillights of the automobile I was following, I might not have been able to keep in my lane. In fact, many cars pulled over onto the shoulder and waited— especially under overpasses, perhaps to protect their hoods and roofs in the event that large hailstones were in the offing. The hot, humid day had certainly prepared the atmosphere for such a fierce occurrence. Fortunately, the rain let up just as we reached our exit to Portland's local streets.

The forty-mile drive to Portland was remarkable for a number of reasons. As far as I could tell, the heavy rush-hour traffic negotiated the sometimes near-zero-visibility thunderstorm without a single accident. Traffic did slow down significantly at times, but by and large it contin- ued to move. I attribute the safe passage of us all at least in part to the skill and caution of the drivers adjusting their speed and patience to the adverse conditions. Their vehicles also helped, being well marked with highly visible head- and taillights, as well as with efficient windshield wipers, good tires that gripped through the water to the road surface, and antiskid braking systems. But, even more importantly, it was the technology of the road itself that enabled us to reach our destination safely.

INTERSTATE 295, THE HIGHWAY we took between Brunswick and Portland, had been resurfaced a few years earlier, and for some time after that work was done the deep-black-colored asphalt road was as free of bumps and potholes as any I know. In good weather, it was a pleasure to drive along the road cut through granite outcroppings dripping with spring water and to enjoy the expansive views of Casco Bay with its many islands and marshy inlets. In bad weather, knowing that there was a solid road beneath whatever water, ice, or snow that was covering it was reassuring. What also made this road so nearly perfect was its engineering design, in conformity with interstate standards that apply throughout America, making driving even on an unfamiliar highway a relatively familiar experience. Its curves, grades, shoulders, and ramps are gentle, gradual, and generous, not only allowing for safe high-speed driving but also allowing for a margin of error for the distracted or drowsing driver by means of rumble strips located just outside the lanes beside the shoulder and median. Exits are clearly marked well ahead with highly readable signs that give ample warning to change lanes if necessary and prepare to leave the main road.

If I have a complaint about Maine's I-295 and highways in general, it is that their lanes and edges are not always marked clearly enough. In driving through the thunderstorm, I found the broken white lines separating lanes and the solid white line dividing the right-hand lane from the shoulder to be barely visible beneath the veil of reflective water washing over them. Were it not for the head- and taillights of the surrounding cars, I might have pulled over and joined the stopped cars waiting out the storm.

The visibility of highway lines is a frustration I have had with many a road, in good weather and bad, in day and nighttime conditions, especially where the lines have become all but totally erased by the rubber tires of traffic relentlessly driving across them. This is not a new problem, and as long ago as the 1930s the California Department of Transportation—familiarly known as Caltrans—began looking for improved ways to mark highway pavement. However, it was not until the greatly increased traffic and traffic accidents brought about by the postwar boom that serious research began under the direction of the Caltrans' engineer Elbert Dysart Botts. The result was slightly raised round pavement markers that came to be known as Botts' dots. When these white ceramic objects were used in place of or in addition to painted lane markings, they not only projected out of standing water on the road but could also be felt and heard when a driver's tires ran over the line they formed. The problem with Botts' dots was that they could be used reliably only in regions of temperate climate, for snowplow blades tended to slice the markers right off the pavement. A variety of raised and depressed reflective lane markings has since been developed, but problems with keeping them undamaged, uncovered, and in place remain.

The visibility of pavement markings has thus continued to be problematic, especially where winter and snow are virtually synonymous. Where salt is used to keep roads clear of snow and ice, its whitish residue masks painted lane markings. During the winter of 2014, the Wisconsin Department of Transportation tested an orange-colored reflective epoxy paint at several locations around the Milwaukee Zoo interstate highway interchange to determine if that color was more readily visible than the federally mandated white and yellow pavement markings even under light snow accumulation and in dark and rainy conditions. Drivers did find the bright orange an improvement, but the results of the test remained to be reviewed by the Federal Highway Administration, which tends to move slowly.

If I were the secretary of transportation, I would make it a research-and-development priority to come up with lane marking systems that remain visible in all weather and light conditions and do not deteriorate under traffic or get ripped up by snowplow blades. And until a lasting solution could be found, I would insist that the left and right edges of the road be repainted regularly with solid yellow and solid white lines, respectively, along with the broken white lines delineating lanes in between—at least as long as those remained the standard colors.

WHICH CAME FIRST, THE yellow or the white? To answer this existential question, we must also deal with a more transgressional one: Why did the driver cross the center of the road? As late as 1917, pavements on rural highways were unadorned with lines or stripes of any kind. It was supposed to be understood that an automobile was to hug the right side of the road when encountering another coming the other way. However, when approaching a tight curve to the left, some drivers were accustomed to cutting the corner and so hugged the left side as they rounded it. This was of little consequence when traffic was slow and sparse, of course, but if a fast-moving vehicle coming the other way was hugging the inside of the same curve, the vehicles would encounter each other in a sudden game of chicken that would leave little time to escape a collision.

Before there were standardized road signs, especially dangerous curves were marked by local residents in an ad hoc way. A "horror sign" might have featured the image of the grim reaper and the words "Just Around Curve" or a skull and crossbones and the warning "Danger— Go Slo." Such notices were expected to serve as adequate reminders and warnings to the driver to proceed with caution.

Marking the road to keep moving vehicles from encroaching upon one another's space is an old idea. About five hundred years ago, a road near Mexico City had its centerline defined by stones of a lighter color than the rest of the pavement. This practice persists in some European cities to this day. In the late nineteenth century, bridges on which a collision would likely do damage not only to the vehicles involved but also to the bridge structure itself had lines painted on their road surfaces to lower the risk. It was also important to control automobiles at crossroads and cross streets and to separate motor vehicles and pedestrians. As early as 1907, stop lines were painted on roads in Portsmouth, Virginia. The first crosswalks are believed to have been painted in 1911 on the streets of New York City.

The American reinvention of the highway centerline, designed to remind motorists to keep to their side of the road, dates from the early twentieth century. But decades before it appeared on an actual American highway, a white line down the center of the road was described in the 1883 utopian novel *The Diothas; or, A Far Look Ahead*, in which were described electric cars capable of traveling at the dizzying speed of 20 miles per hour on the open road. During a demonstration ride, "the white line running along the centre of the road" was explained to the narrator: "The rule of the road requires that line to be kept on the left, except when passing a vehicle in front."

For real roads in real time, the concept of the white centerline is commonly attributed to Edward N. Hines, a charter member of the Wayne County, Michigan, Road Commission, on which he served from 1906 until his death in 1938. It is Hines's idea that was once described in the trade magazine *Michigan Roads & Construction* as "the most important single traffic safety device in the history of auto transportation." Among Hines's other notable achievements was instituting the construction of that first mile of concrete roadway in Detroit in 1909. Two years later, he is said to have observed a milk wagon leaking some of its contents and leaving a white stripe behind it as it progressed down the road. This is said to have given Hines the idea for the modern centerline, the first of which was painted on River Road in Trenton, Michigan, just south of Detroit.

The first centerline on a rural highway is believed to have been painted in 1917 in milk white on the portion of the Marquette–Negaunee Road known as Dead Man's Curve, which is located on Michigan's Upper Peninsula. The line was the work of Kenneth Ingalls Sawyer, who was veteran superintendent of the county board of road commissioners and had drafted the state's first gas and weight tax laws, along with a good deal of its other basic highway legislation. As he wrote in the September 1920 issue of *Municipal and County Engineering*, "the handling of motor traffic upon our main trunk highways through the country is rapidly becoming as serious a problem

as traffic control has ever been in our cities." It was, therefore, necessary
to adopt methods of urban traffic control in rural areas. Sawyer related
how traffic between the towns of Marquette and Ishpeming, farther
down the road past Negaunee, had "become heavy enough to make
travel dangerous unless some means of control is adopted." His means
was "white 8 in. center lines upon the black surface of the road upon the
more dangerous curves, with an arrow pointing down the right hand
side of the road at either end." He believed that drivers responded to the
white line because they had become accustomed to obeying similar
devices controlling traffic in cities. The result was an "immediate reduction in the number of accidents." Sawyer acknowledged, however, that
the centerline was not a panacea; it worked on that road because of its
"smooth black surface," which allowed the white line to "stand out in
sharp relief." To maintain that condition of visibility, the highway
patrolman touched up the line every Saturday morning, something not
practical on today's interstates.

*The first known highway centerline is believed to have been hand-
painted in 1917 around Dead Man's Curve on the road between
Marquette and Ishpeming, on Michigan's Upper Peninsula. White
paint was used, and the arrow was intended to remind drivers on
which side of the road they were to remain, something that was not
yet second nature to all motorists of the time.*

Clearly the goal of road safety would be better served if the design of road signs and markings did not vary from urban to rural roads, from state to state highways. Following studies, conferences, and reports, the American Association of State Highway Officials and the National Conference on Street and Highway Safety issued in the late 1920s manuals for signs and control devices on rural and urban roads and streets. But two separate manuals did not help standardization, and so in 1932 a Joint Committee on Uniform Traffic Control Devices was formed, and in 1935 it published in mimeographed form the first *Manual on Uniform Traffic Control Devices*, known as *MUTCD*, which soon became an American standard. The great demand for the manual caused a printed version to be issued in 1937; the 166-page document covered signs, markings, signals, and islands. (The latest edition, published in 2009, is 864 pages long.)

The first manual did not require centerlines everywhere. They needed to be painted only on hill-crest approaches with limited sight lines; tight or restricted-view curves; and pavements wider than forty feet. The acceptable colors for the lines were white, yellow, or black, the choice being dependent on which provided the greatest contrast against the pavement, whose color could vary from coal black for asphalt to almost white for concrete. (Such considerations remain important and dictate that lane markings on concrete-surfaced bridges consist of white lines painted over wider and longer black ones. Concrete taxiways and runways at airports are typically marked with yellow or white lines outlined in black.)

By 1954, forty-seven of the then forty-eight states had adopted white as the standard color for the highway centerline. The holdout was Oregon, whose highway department believed yellow to be the more visible color. Indeed, it is easier to see under a dusting of newly fallen snow, but the State Highway Commission capitulated. In an editorial on the matter, the *Oregonian* newspaper incidentally noted that the claim that the highway centerline originated in Oregon was unsubstantiated, attributing the concept to "several states" that had "hit on the idea independently at about the same time." However, in a letter to the editor, Peter V. Rexford, a retired captain of the Multnomah County

Sheriff's Office, claimed that "the first yellow centerline ever painted on pavement" was done under his direction in April 1917. He may have been correct that it was the first yellow line.

Oregon citizens were not happy that at midcentury their traditional yellow road markings had been changed, and public pressure caused the highway commission to reverse itself. Thus after two years of trying out white, yellow was reinstated. And then just two years later, in 1958, the Bureau of Public Roads decided on white as the standard for lines marking the new interstate highway system. If Oregon did not conform, it stood to lose at least $300 million in federal funds. Citing a series of tests that found yellow to be the safest color deer hunters could wear, the *Oregonian* editors argued for keeping yellow lines on roads under the sole jurisdiction of the State Highway Department. As for the interstates and U.S. highways, the yellow lines would have to be changed to white.

Soon the federal government reversed itself. The 1971 edition of the *MUTCD* declared a new standard for marking two-way roads, whether containing a median or not. To emphasize that they separated traffic moving in opposite directions, all centerlines were to be painted yellow. Where there was a median, the line marking the left edge of the leftmost of two or more lanes was to be yellow. White was to be reserved for regulating traffic moving in the same direction. Thus, broken white lines would continue to separate adjacent lanes carrying traffic in the same direction. Oregon had to change its roadway colors again, but this time it did so gladly.

As IMPORTANT AS KEEPING vehicles from crossing over the centerline is, preventing them from veering off the side of the road is equally imperative. The solid white line that marks the right-hand edge is thus considered another significant innovation. Although the idea may have been on the minds of many in the early 1950s, it was the chemist and metallurgist John Van Nostrand Dorr who was prompted by his wife to do more than talk about it. The Dorrs, like a lot of nighttime drivers, tended to hug the centerline, which aggravated the glare of approaching headlights and caused the driver to veer to the right toward the shoulder.

Dorr convinced the Connecticut highway department to paint an experimental right-edge line on a four-mile stretch of the Merritt Parkway. When the line proved to help drivers stay centered in their lane, the entire parkway was so painted, along with many additional miles of the state's busiest roads. Other states soon followed.

Whereas wheeled contraptions attributable to numerous Rube Goldbergs were used to paint highway centerlines as early as the 1920s, the invention of an edge-line machine is attributed to one John Edward White, who worked for the Ohio Department of Transportation. One day, while driving down a highway in dense fog, he was having difficulty seeing the road. To drive on, he had to keep his head out the window to watch the median line. Apparently independent of Dorr, he conceived of painting a white line on the right side of the road to mark its edge. In dense fog a navigator might have to stick his head out the passenger window, but that would be safer than the driver exposing his head to oncoming traffic. In 1956, to implement his idea more efficiently, White developed a prototype edge-line machine employing a subcompact Crosley automobile chassis. He was encouraged to do this because, whereas the 1948 manual had recommended against the use of edge markings generally, a 1954 revision modified the prohibition. The white line delineating the pavement's right-hand edge was explicitly advocated in the 1961 manual, and the 1978 edition made it required for all multilane rural highways.

It had been a milestone in road marking to mount line-painting equipment on a maintenance vehicle, thus eliminating the need for the manual application of paint with a brush. Not only could the lines be laid down faster, but the job could be done more safely as well. Perhaps even more important than the method of application, however, was how long it took the paint to dry. As long as it remained wet, vehicles running over it would track the paint across lanes and create a distracting and confusing pattern. And as long as paints were slow drying, traffic cones had to be placed along a newly painted line to keep cars and trucks off of it. This was, of course, labor-intensive and dangerous for workers and frustrating for drivers. Faster-drying paints and techniques to apply heat to speed up the drying of the freshly painted line were eventually developed. This meant that a string of maintenance trucks, which might include one sweeping off the road surface, one laying down the paint, one applying heat to it, and one bringing up the rear to warn of and protect the slow-moving train of trucks, could move at such a pace that the lines could be driven over almost as soon as they were exposed behind the last truck.

Another important milestone was the addition of glass beads to the paint itself to make it "retroreflective," meaning that light from a source like a headlight would be reflected back, thereby making the surface in which the beads were embedded more visible in nighttime driving conditions. One of the pieces of equipment in a line-painting train of vehicles would apply such beads to the wet paint; clearly the nature of the paint and beads and the timing of their application would be critical to how effectively they worked. Retroreflective tape and sheeting, in which beads and similar small reflective particles are already embedded, can also be used on pavements and signs to make lines, crosswalks, and directions more visible at night. Of course, the use of beadlike devices, no matter how small, can also make road markings more vulnerable to the action of traffic and, like Botts' dots and their descendants, snowplow blades. Today, some reflective devices used to mark traffic lanes are installed in depressions in the pavement to put them out of reach of plows. Technological advancements in road safety continue in parallel with the development of more refined standards.

WHEN THE 1971 EDITION of the *Manual on Uniform Traffic Control Devices* specified that highway guide signs—those that identify exits, for example—be made up of white text on a green background, over a half century of debate over the shape, content, and color of street and highway signs generally ended—at least until some future edition of the manual. Before the 1920s, roads and their signage were unregulated. Private automobile clubs tended to install signs to promote a highway and the businesses located along it rather than to help the traveler get safely from point A to point B. Competition among clubs could result in as many as eleven different sets of signs on a single highway. The first organized attempt to standardize road signs took place in the early 1920s. After touring several states, a group of representatives from Indiana, Minnesota, and Wisconsin reported their findings to the Mississippi Valley Association of Highway Departments, which in time adopted, among other things, the suggestion that different kinds of signs have different shapes, including circular to warn of a railroad crossing, diamond-shaped to signal caution, and octagonal for a stop sign. Even when its face was poorly illuminated, the shape of a sign could provide valuable information.

The evolution of the stop sign is a multifaceted story. An early version of such a sign was proposed as early as 1900 by William Phelps Eno, who was born in 1858 in New York City to a family whose wealth derived from investments in real estate. Growing up in an age when horses were a principal means of transport, he experienced numerous situations of horse-and-wagon gridlock. On one memorable occasion, seven-year-old William and his mother were being driven home in an open carriage known as a barouche to the family brownstone mansion on East Twenty-Third Street. They were coming back from a visit to his father's office on Lower Broadway when they encountered a traffic jam that took a half hour to untangle. It made a lasting impression on the young boy. As he would later recall, "It seemed as though a dozen vehicles could cause a blockade, since neither drivers nor police, if there were any around, knew anything about the control of traffic or the proper thing to do." When he and his family traveled to Europe, he observed traffic in Paris, London, and other cities; traffic and its problems became his preoccupation. It became a goal bordering on obsession for Eno to figure out what to do about traffic and its control.

While still a young man, Eno was repeatedly frustrated by the congestion that resulted when the opera or theater let out, and he eventually devised a system of carriage dispatching to alleviate it. He would not become an engineer by formal training—he studied architecture at Yale—but he would essentially become one by temperament and practice. Although he worked on real estate matters in his father's office from 1884 until 1898, his mind kept wandering back to traffic. At middle age he retired from the business world to spend more time studying traffic problems and formulating solutions, eventually to become known as the "father" of traffic regulation and safety. In 1900 he published his first article on traffic movement in *Rider and Driver* magazine under the title,

From his childhood growing up in New York City, William Phelps Eno was fascinated by traffic and its congestion, which he took as a problem to be solved. Without any formal training in engineering, he worked independently throughout his life on schemes to control traffic, publishing groundbreaking books on the subject and becoming known as the father of traffic regulation.

"Reform in Our Street Traffic Most Urgently Needed." In 1903 he produced the brochure *Rules for Driving*, which formed the basis for New York City's (and the nation's) first written traffic code. For the purposes of his rules—the software of street and highway infrastructure, really—he defined the word "vehicle" to include "equestrians and everything on wheels or runners, except street cars and baby carriages."

In that same year Eno had made and posted the first traffic control signs, which directed slow-moving vehicles to drive near the right-hand curb and, in smaller text, informed that "Rules for Driving Can be Obtained at All Police Stations." The 10-by-19-inch enamel signs had white lettering on a blue background, and one hundred of them were

This is believed to be what the very first formal traffic sign looked like when posted in 1903 in New York City. The blue-and-white enamel sign was created and paid for by William Phelps Eno, a well-to-do New Yorker who became obsessed with solving problems of traffic congestion on city streets. He was also the author of Rules for Driving, *which he had printed and distributed. Among the rules Eno codified was driving on the right side of the street.*

erected. According to Eno's own assessment, which he expressed in his 1909 book, *Street Traffic Regulation*, the signs were effective but were "rather small." The following year he had similar but larger signs made. However, these were "too large, poorly made and badly hung, most of them being too low and in the way." Eno was nothing if not critical, even of his own work. He concluded that the "proper size" was about 20 percent larger and with about 50 percent more words than the originals. By today's standards, they would be considered too wordy.

On the occasion of the publication of *Street Traffic Regulation*, the *New York Times* noted that it was "due largely to Mr. Eno's efforts that the handling of the street traffic of New York has reached its present satisfactory condition." It was not easy, however, for as he explained in the book's preface he had to overcome "the indifference and ignorance of city officials, and their slowness in realizing their duties" regarding the new concept of traffic control. Indeed, Eno's book was the first on the topic. Among the organizing rules the self-taught traffic engineer laid down was driving on the right side of the street and giving northerly and southerly moving traffic in Manhattan right of way over that moving easterly and westerly across the island.

Over the course of his amateur career, Eno published a number of books on traffic regulation, including one in French, for his studies and influence were international. The early books were produced by him at his own expense and distributed freely. One recipient of his first book, the 1909 volume *Street Traffic Regulation*, wrote of it in a note of thanks alluding to the Victorian poet Robert Browning that "weighed against Browning's works, for example, I think it is fair to say that it will do much more good in the world."

Eno's work lives on in the Eno Transportation Foundation, originally the Eno Foundation for Highway Traffic Regulation, whose seal bore the Latin motto *Ex chao ordo*, which translates as "Out of chaos, order." The foundation was established in 1921 and headquartered in the Eno family's thirty-room summer home at Saugatuck, Connecticut, now part of Westport, located on Long Island Sound. In 1938 a separate headquarters building was constructed nearby. Today the foundation's Eno Center for Transportation is a think tank located in Washington, D.C. Since 1924, the foundation has been issuing studies and reports relating to traffic and transportation issues ranging from matters of parking, speed, and regulation to administration and management of transportation organizations. The motor vehicle may have replaced the horse since Eno's first forays into traffic rules and regulations, but the man remains known for his innovations, which include, in addition to driving rules, the taxi stand, the traffic island, and the stop sign.

The early stop sign—and the action it commanded—was known as a "boulevard stop," because it was placed on streets crossing the major roads that often led radially out from the center of a city to its suburbs. A system of these "park boulevards"—also known as boulevard stop streets and traffic ways—was introduced first in Chicago, and adoption followed in such cities as Detroit, Los Angeles, and New Orleans. As a reminder of the regulation, some cities painted notice on a cross street that a driver was approaching a boulevard stop. A line indicating where to stop was also painted on the street. However, without a physical stop sign installed at the intersection, the rule was often ignored. Neighborhoods through which the boulevards ran opposed the installation of the stop signs, which they claimed sped up traffic having the right of way. Not only did this make it more difficult for vehicles and pedestrians to cross the major artery, but also it encouraged more traffic on it, thereby favoring the suburban over the urban resident. Furthermore, traffic surveys indicated that in some cities, where as many as 98 percent of vehicles complied with a traffic light, more than half of the vehicles using cross streets did not make a full stop at a sign. Stop signs elicited such strong feelings of opposition to them that in 1922 they were ruled illegal by courts in Illinois because they were "a violation of the right of individuals to cross streets."

Early signs instructing vehicles to come to a full stop before proceeding across or turning onto a street with fast-moving traffic were known as boulevard stops. This diamond-shaped example from Southern California was put in place before the octagon shape came into standard usage. California consistently used white letters on a red background for its stop signs, even during periods when the American Association of State Highway Officials advocated yellow as a background color.

Early stop signs were—like other signs signaling caution—diamond shaped. The now-familiar octagonal shape of the stop sign is attributed to Harry Jackson, a Detroit police sergeant who cut off the corners of a square- or diamond-shaped sign to produce the octagon. Initially it, like most other signs of the times, had black letters on a white background. However, in 1924, as an outcome of the First National Conference on Street and Highway Safety, different colors were recommended to further distinguish traffic signs. For the stop sign, the white letters were to be set against a red background. Around the same time, the American Association of State Highway Officials issued a report recommending octagon-shaped stop signs with yellow as a background color, since red faded and was not as visible under nighttime conditions on rural roads. A manual for signs on urban streets preferred red letters on a yellow background for the stop sign. The difference was not fully reconciled until fade-resistant finishes were developed: the 1954 supplement to the 1948 manual specified that the word STOP should appear in white letters on a red background. According to the William B. Kolender San Diego Sheriff's Museum, California is the only state always to have used red.

THE EVOLUTION OF THE traffic light was no more direct than that of the stop sign. Although railroads had standardized track signals well before the automobile appeared on city streets, the obvious move of adopting the railroad's practices was not simply taken over to control street traffic, for the challenge seemed more complicated than telling a locomotive engineer whether the track ahead was clear or not. Unlike a railroad train that could travel for great distances without encountering an intersecting track, it was the nature of city-street traffic to come upon cross traffic every few hundred feet at most. According to the urban and social historian Clay McShane, traffic signals evolved in response to motor vehicle traffic jams, which became intolerable diurnal events in the years following the introduction of the affordable Model T, whose left-side placement of the steering wheel was a significant factor in firmly establishing the practice of our driving on the right side of

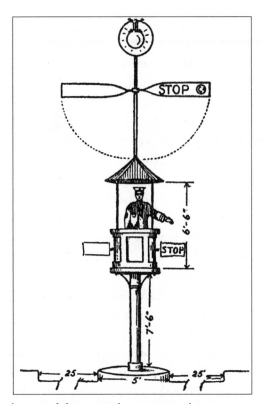

In the early years of the twentieth century, city drivers were accustomed to cutting corners when making a left-hand turn at an intersection. To encourage them to turn around the center of the street crossing, posts and standards—some temporary and some permanent—were installed in the middle of the intersection. With increasing traffic, these evolved into traffic towers incorporating a crow's nest from which a policeman manually directed traffic and operated stop signals.

the road. Police officers were generally ineffective in controlling high-volume traffic and breaking up the jams that occurred at intersections. Early mechanical devices did owe their inspiration to those used on the railroads (which in turn derived from maritime signals and those in turn from lighthouses). They included semaphores and signal towers, but on city streets they typically required a policeman to operate them manually.

Not all signal-operating policemen were safe in traffic towers. According to AASHTO's centennial retrospective, it was a pavement-based policeman in Salt Lake City who "established the foundation for traffic signals" as we know them today:

> Officer Lester F. Wire didn't feel safe when he was directing cars, trucks, buggies, and trolleys at a busy intersection in downtown Salt Lake City in 1912. So he built a birdhouse made of plywood, painted it yellow, and punched six-inch holes on either side. He then dipped bulbs in red and green paint and used a manual switch to change the lights from red to green. By 1917, Salt Lake City had traffic signals at six connected intersections, all controlled simultaneously from one manual switch. It was the first interconnected traffic signal system in the United States.

In a book of photographs documenting block by block New York's Fifth Avenue from Washington Square Park to East Ninety-Third Street as it was in 1911, the compiler noted that "there is nary a traffic light nor a sign to be seen in any of the photographs." But the time was ripe for change. Cleveland had a permanent installation of a traffic signal in 1914. It had only two functioning faces, whose red and green lights controlled two-way traffic on the main street only. Police officers supplemented the signal by directing traffic coming out of the cross streets. The presence of the officers was thought to be needed also to make drivers and pedestrians obey the signal lights, for otherwise their commands might go unheeded. Another problem with some early traffic signals was that even when they were fitted with four faces, they still had only red and green lights, which meant that the change from go to stop was more sudden than drivers could respond to, so they continued through the intersection for some seconds after their light had turned red and the cross street's green.

Early traffic signals were employed not only to control ordinary street traffic but also to warn such traffic when speeding fire or police vehicles were responding to a call. This was one of the functions of the "Municipal Traffic-Control System" for which Cleveland inventor James B. Hoge received a patent in 1918. As he noted in the patent document, the warning sounds of approaching emergency vehicles might not be heard easily among the ambient sounds along busy city streets, and so visual warning signals were to be preferred. Hoge's extensive system of electrically interconnected street-corner signals involved backlit signs reading STOP and MOVE. Thus, the synchronized system relied more on words than colors to communicate with drivers.

Another solution to the problem of controlling traffic was invented by another Cleveland resident, Garrett A. Morgan, who, based on his 1923 patent for an improvement in a kind of traffic signal, is often incorrectly credited with the invention of the traffic light itself. Morgan's invention, for which he filed for his patent in 1922, was in fact a manually operated semaphore device whose movable arms could be raised to display a STOP signal in all four directions at an intersection, thereby halting all traffic briefly before being turned to display a GO signal for one of the streets. For night visibility, the words were illuminated by a pair of bulbs, but like Hoge's STOP and MOVE system the device was regressive in the sense that it relied on words rather than colored lights to signal a command. Nevertheless, it did lower the probability of a car running a red light, which even today is the cause of many an accident.

In traffic signals that relied exclusively on different-colored lenses, the dangerous situation was ameliorated somewhat by allowing the green light to remain illuminated for a few seconds after the red came on so that oncoming drivers were given warning that the light was changing. At the same time, the change from red to green for the cross-street traffic was delayed. The addition of the now-familiar yellow caution light occurred in 1917 in Detroit. The concept is credited to city police officer William Potts, who in 1920 was also responsible for the first four-faced three-colored light. However, early versions of this improvement had only three lightbulbs internally, one each illuminating separately the four top, four center, and four bottom lenses. This meant that whereas on two opposite faces of the signal the order of the colors could be, from bottom to top, green, yellow, red, on the other two faces the order had to be red, yellow, green.

In 1917, as an experiment, New York City installed a primitive traffic signal at the intersection of Fifth Avenue and Fifty-Seventh Street. It was the invention of an engineer named Foster Milliken and consisted of something like a flashlight that could be directed at oncoming vehicles and flash red or green to signal stop or go. But not all early traffic signals adopted those colors. As late as 1924, a citizen observed that "our signal system provides an orange signal for go, a green signal really for stop . . . , and a red signal which may mean stop, but is actually taken as a 'getaway' signal." Drivers must have been more confused than controlled.

The Fifth Avenue and Fifty-Seventh Street experiment was short-lived. A true modern traffic control system consists of not just one signal at a single intersection but a series of coordinated signals located at key intersections along a major street or avenue. This was the idea of wealthy New York physician John A. Harriss, who was irresistibly drawn to the study and improvement of street conditions. Like William Eno, Harriss worked out his own solutions to unsnarling city traffic and spent his own money implementing his ideas. In 1920, when he was the city's deputy police commissioner in charge of the traffic division, he erected along Fifth Avenue a series of five traffic towers, each located at a key intersection. Each tower was little more than a small wooden shed elevated about twelve feet off the street on spindly steel legs. Each Harriss signal tower was occupied by a policeman who, in coordination with other tower operators, manually changed the signals. According to one source, a green light meant go for the traffic on the avenue, a red meant stop, and a flashing white (or amber or yellow) light meant the signal was going to change. However, another source reports that green meant stop and white go. There seems to be a good deal of confusion among chroniclers of the period about what color signaled what, which was perhaps not so surprising for observers of a new system.

Though traffic was not totally unjammed, under Harriss's system it did move a bit more quickly from Thirty-Fourth to Fifty-Ninth Street. Before the traffic towers were erected, that mile-and-a-quarter trip took forty-two minutes. With the towers in place, it took nine minutes. Thus the system was effective, but the towers were unsightly. So the Fifth

Avenue Association, an organization that took an active interest in preserving the character of the famed street and its environs, sponsored a national design competition for a more appropriate traffic signal tower. From the 130 entries, that of the well-known New York architect Joseph H. Freedlander was chosen. His art deco structure with neoclassical ornamentation consisted of a twenty-four-foot-tall tower made of cast bronze over a steel frame that sat on a five-foot-square concrete base. As installed, the structure was surrounded by a traffic island that was offset from the center of the intersection so as not to interfere with cross-street traffic.

The traffic towers were operated on the so-called continuous block system, in which all of them changed from stop to go simultaneously. William Eno was opposed, based on his scientific observation and analysis, to the use of such a system on city streets. He told the *New York Times*, which identified him as "one of the earliest students of traffic conditions," that the block system "wastes too much of the traffic capacity of the street because between the time the signal is given to stop and the time the signal is given to go the vehicles have gone ahead a long distance from the place where they were first stopped to the place where they are stopped again. This leaves a large portion of the street surface unoccupied by vehicles, all of which is pure waste of street surface." He suggested that if the avenue were photographed from an airplane when all the traffic was waiting for the red lights to change, most of the street surface would be vacant. The practice would drive through traffic off the avenue and, he argued, leave more room for the clients of Fifth Avenue tenants, thereby increasing the profits of their stockholders. Enacting traffic rules that resulted in this kind of disparity he considered "class legislation" that was "good for the few but bad for the general interest of the city."

Eno's ideal was to have every square foot of road surface occupied by vehicles that were moving in perfect order at all times. In lieu of traffic signals, he advocated the use of a rotary system, which he believed added as much as half again to the capacity of a street. In addition, a system of roundabouts would "equalize the speed of vehicles to a safe mean." Eno was responsible for the redesign of the traffic pattern

Early traffic signals on New York's Fifth Avenue consisted of towers erected in the middle of the street. Each tower was manned by a policeman who observed the traffic from an enclosed crow's nest and changed the signals manually. Architect-designed towers made of steel, cast bronze, and glass were installed in the early 1920s but were removed by the end of the decade in favor of traffic lights on posts located on sidewalk corners.

around New York's Columbus Circle. Traffic had been running two ways around the large center island, and accidents had been common. He suggested changing the two-way traffic pattern to one-way, and when that was instituted in 1905 it did solve the problem. Indeed, the new traffic pattern around Columbus Circle served as a model for that around London's Piccadilly Circus and around the enormous rotary at Paris's Arc de Triomphe.

As the 1920s roared on, the traffic towers sitting in the middle of Fifth Avenue became intolerable obstructions, and they were removed in 1929. Their architect was asked to design "a simple bronze lightpost" that could stand on a sidewalk corner of an intersection. Joseph Freedlander rose to the occasion and produced a streamlined signal that contained just red and green lights and was topped by a seventeen-inch-tall bronze figure of Mercury that was finished in gold leaf. Although a joint board of federal and state highway officials had agreed in 1925 upon the standard red-yellow-green arrangement so familiar today, the intermediate yellow light was outlawed in New York, because drivers had been taking its blinking announcing a change to green to be license to proceed prematurely. In my experience, New York drivers have maintained a habit of pushing the limits with traffic signals. When the yellow reappeared, it was with a different function. Unfortunately, rather than taking it as a signal to brake, drivers took it as license to race to beat the red light. Indeed, even though not a New Yorker, one of our neighbors is fond of accelerating in order to "squeeze the lemon" as she tries to clear the intersection before the tomato appears.

Not until 1928 were three-color traffic signals fitted with twelve bulbs so that the red light could appear on top on all four faces. This order of red on top became the standard in 1930, which was a boon for the 10 percent of male drivers who are color-blind and see both red and green as gray. This was no consolation to the Irish-American community in Syracuse, New York, who wished traffic lights in its neighborhood to display green on top, a home-country bias that persisted into the 1960s. In their failure to settle on a uniform use of color in signals—not to

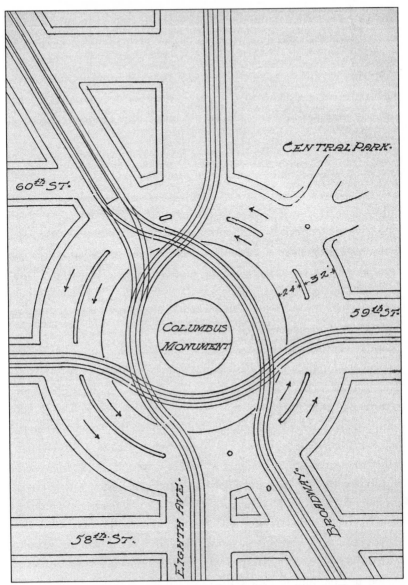

William Phelps Eno accepted the challenge of rationalizing traffic flow around Columbus Circle in New York City. His recommendation was to make all traffic go in the same direction—counterclockwise—and to install "isles of safety"—traffic islands—to which pedestrians could progress as they attempted to cross multiple lanes of moving motor vehicles, horse-drawn wagons and carriages, and trolley cars. The plan was instituted successfully in 1905.

mention their decades of deliberation over the choice of yellow or red for the background color of early stop signs, or their vacillation over white or yellow lines on the pavement—traffic engineers may have left the impression that they had considerable difficulty in making choices and settling on standards. But there may have been good reason for such deliberation.

Red was long considered a color generally associated with danger, and so choosing it to signal stop made sense. However, there was ambiguity in some of its uses, such as for exit signs in theaters. At a 1922 meeting of the American Railway Association, one engineer noted that exit signs marked routes to safety in cases of fire or other emergency and so should be illuminated in a color other than red. Even the color of automobile taillights was questioned, with many engineers thinking yellow to be preferable to red, perhaps to signal caution. The use and significance of color remains somewhat ambiguous today. The next time you are on an interstate highway, look at the array of fast-food-restaurant logos on the food-fuel-lodging signs preceding exits. For fast-food restaurants especially, red is definitely the predominant color used, often in combination with yellow. Think McDonald's, Burger King, Wendy's, Pizza Hut. Someone—more likely a team consisting of a color consultant, a psychologist, a marketing expert, and others—advised on the color scheme, and the corporate decision makers were convinced by the argument. Today, as they do on traffic lights, yellow and red mean respectively prepare to stop and stop, exactly the actions that restaurants wish to elicit from the traveler. For a long time I was confused by the white-and-yellow (with a hint of green) logo of Subway sandwich shops. It stands out as an anomaly among the predominantly red-and-yellow logos, but when I recall the extended battle over the colors of road markings—which did not include red as a choice—I suspect there may have been some serious reasoning behind Subway's thinking. Then there is the Starbucks green logo, without a hint of red or yellow. Green is, of course, the color of environmentalism and responsible stewardship; it is also the color that says go, which is what patrons of a fast-food chain usually want to resume doing shortly after they stop. Perhaps the Subway and Starbucks logos are subliminally signaling drivers that they will not be unduly delayed at those shops.

INDEPENDENT OF THE QUESTION of colors in traffic signals, engineers wrestled with more explicitly technical issues when things did not work as well as they conceivably might. In the case of manually operated signals on Fifth Avenue traffic towers, for example, there was the need for policemen in other towers and on the ground constantly to check what the main tower was signaling so they could be in line with it. Given the human limits of distraction and delayed response time, it was unrealistic to expect perfect synchronization. However, that could be achieved with electrically linked and controlled signal systems. Where they worked, automatically timed traffic lights were adopted with start-ling speed. New York City, for example, which had 98 lights in 1926, added 1,143 in 1927, and another 2,243 in 1928. While the cost of installation was significant, the traffic lights enabled the city to reduce the number of police officers on the traffic squad from six thousand to five hundred, resulting in an annual savings of $12.5 million. However, the more sophisticated automatic controls involved required engineers to design the system within which the signals operated. One early strategy was the block system that William Eno railed against. The downside of such a scheme was that when the lights on the avenue all turned green simultaneously, cars raced to get through as many of them as they could before the red lights appeared.

An alternative was to stagger the lights so that as a motorist drove down a one-way avenue at a speed of, say, 25 miles per hour, he would find each red light turning green just as he approached it. This was the system employed in New York City when I worked for its traffic depart-ment one summer in the early 1960s. The installation of a set of signals at an intersection that had not previously contained them involved as a final procedure determining the timing to fit the new lights into the stagger. It was very satisfying to drive along an avenue so properly controlled that, of the red lights displayed as far as the eye could see, only the closest one turned green just as my car approached it. What worked in theory on large sheets of graph paper in the office also worked in practice in the urban field. It was as joyous to watch as an elaborate piece of clockwork in a Gothic cathedral—or a line of Rockettes performing a serial bow in Radio City Music Hall.

FOR OBVIOUS REASONS, THERE are few traffic lights on interstate highways. The one I know of is on I-95 where the multilane road crosses the Maryland-Virginia state line via the Woodrow Wilson Bridge across the Potomac River. One of its spans is a bascule and hence has to have a light to stop traffic when the leaves of the drawbridge are raised, which is very infrequently. In decades of driving over that bridge, I have yet to be stopped by a red light. Elsewhere on I-95 there are other major spans, but these bridges have a high enough clearance over water that boats and ships are not impeded by their presence. Traffic on the interstates generally keeps moving, of course, unless there is an accident, construction, or just too many vehicles to progress smoothly through transitions from more lanes to fewer.

The stretch of I-295 that we took to Portland—along with other interstates in Maine—is so well designed that the flow of traffic routinely exceeded the 65-mile-per-hour speed limit. There were a few terrible accidents on the road, but the cause of them was often attributed not so much to drivers going a bit over the speed limit but to other drivers adhering to it too rigidly—or, worse, going much slower than the limit— thereby encouraging a lot of changing lanes and passing. In recognition of this phenomenon, just a few months before our excursion to Portland, the Maine Department of Transportation had raised the speed limit to 70 on I-295 and similarly raised it from 55 to 60 on U.S. Route 1, which feeds the interstate spur. In fact, at the time, speed limits were raised on roads throughout the state. The move was not without opposition, of course. For one thing, it did not address the problem of the slowpoke driver, but transportation officials looked upon the change as an experiment that could easily be dialed back if accidents increased rather than decreased. Another counterintuitive change in traffic control once took place in England, where some road markings were removed after it was found that drivers were more cautious when the usual traffic-control devices were absent. Time will tell whether decisions like these will in the long term prove to have been wise or foolish.

Such is the nature of infrastructure. It is roads and bridges and more things that have to be made adaptable to changing conditions of weather and traffic demand; it is expansion and alteration to accommodate changing patterns of industry and settlement; it is men at work constructing the new and maintaining the old; it is lane shifts and crossovers to make way for work crews; it is slowdowns and detours to replace bridge piers and girders; it is anecdote and experience about how to get from here to there quickly and safely. As large and extensive as our nation's interstate highway system is, its users have come to know it as something familiar. They attribute qualities to the highways that go well beyond their number designations. A stretch of expressway can get a reputation as a good road or a bad road, as a speed trap or a parking lot.

In an automobile driver's ideal world, every highway would be newly paved, perfectly marked, and absolutely free of trucks. The road would be as smooth as a mountain lake on a windless day. All drivers would give their full attention to the traffic in front, beside, and behind them. They would not eat a sloppy sandwich, sip a cup of scalding coffee, use their cell phones, listen to overly captivating books on tape, daydream of conducting the orchestra playing on the radio, reach down to the floor to find the quarter for the upcoming toll, look aside to adjust the temperature, look up to stretch their aching neck, look around to scold a child or pet, or close their tired eyes for just a second. The perfect driver is as rare as the perfect road, however. Tractor-trailer trucks running with retreaded tires can leave debris even on a perfectly smooth road, causing drivers to swerve to avoid it. Cars and trucks drop parts of themselves—nuts, bolts, tailpipes, mufflers, fan belts—which also can become distractions or worse and lead to accidents. And this does not only happen on highway infrastructure. Recall that it was a piece of metal from an airplane taking off previously that a supersonic Concorde ran over on a Paris airport's runway, causing the object to puncture a fuel tank and ending up with the Concorde bursting into flames and crashing.

Infrastructure does not take care of itself. No matter how well designed and built, a runway or highway must be kept smooth and fit for its purpose. Left unattended to, cracks grow into gaps, gaps into holes, holes into bumps, and big enough bumps into trouble. This is why rights-of-way must be watched and lanes closed for maintenance every now and then. Infrastructure demands vigilance. The tarmac, the taxiway, the ramp; the underpass, the overpass, the bridge; the service road, the sidewalk, the curb; the fireplug, the lamppost, the tree—each can be both friend and foe, depending on the time of day and a pilot or driver's state of alertness or drowsiness. Each component of infrastructure can be as variable as the weather. The details make the difference between technological success and failure, whether in the design of airplanes or automobiles or the infrastructure over which they travel. Some of the most important details of interstate highway design are so inconspicuous and subtle that we hardly notice them, at least consciously, yet it is they that can make the difference between having a safe journey and having a fatal accident, especially in inclement weather.

6

Could Not Travel

EARTHQUAKES, ENVY, AND EMBARRASSMENTS

A LOT OF ROAD building took place between the time when young Lieutenant Colonel Eisenhower observed the sorry state of American infrastructure while riding in the Army's transcontinental convoy and when he was president and signed the law establishing the interstate highway system. Major state turnpikes built in the interim were incorporated into the new system, and their toll-collecting practices were grandfathered. In the meantime, major bridges had also been constructed, and they, too, were incorporated into the system. In California today, for example, Interstate 80 crosses San Francisco Bay on a bridge that was completed in 1936 as one of the greatest engineering achievements of its time. Seventy-five years later the replacement of part of this bridge became one of the nation's great engineering embarrassments and one of its most infamous infrastructure case histories.

The city of San Francisco and the Golden Gate Bridge are inextricably linked in the minds of tourists, but to many a commuter into the city from the east side of the bay the more significant fixed crossing in the area is the San Francisco-Oakland Bay Bridge, popularly known as the Bay Bridge. It was this double-decker structure that was visibly damaged in 1989 when the Loma Prieta earthquake caused a fifty-foot section of the upper roadway to fall onto the lower. A motorist who drove into the gap was killed, and the entire structure had to be closed to traffic while the roadway was being repaired. The loss of the bridge emphasized the obvious: that the crossing is a vital link in the area's infrastructure chain upon which the regional economy relies greatly. Uncommonly rapid repairs reopened the bridge in about a month, but damage and risk assessments made it clear that the entire structure had to be updated to make it more earthquake-resistant. Thus began a saga of design and construction that took more than two decades to play out, and even then without full resolution. In the meantime, because of the attention brought upon the structure, the history and significance of the Bay Bridge has rightfully regained a place of distinction in both the public and professional memory.

Talk of a bridge across the miles-wide San Francisco Bay dates back at least to the middle of the nineteenth century, when suspension bridges with a main span exceeding one thousand feet had become reality. In 1849, one such bridge was completed over the Ohio River at Wheeling, West Virginia, and another over that same river was soon under construction at Cincinnati. Proposals to span even greater distances were being bandied about in New York, Philadelphia, St. Louis, and elsewhere, including the San Francisco Bay area, where it had become a topic of discussion even among individuals who were neither engineers nor architects.

The gold rush attracted a lot of fortune seekers to the West Coast, and Joshua Abraham Norton proved to be one of the more colorful. Norton was a Scottish businessman who struck it rich in California—and then lost it all, including his grip on reality. He declared himself Emperor of the United States, Protector of Mexico, and Sole Owner of the Guano Islands. These Pacific islands contained mounds of seabird excrement, which represented a valuable source of fertilizer. The Guano Islands Act of 1856 enabled U.S. citizens to take possession of any such island—regardless of where located—not already occupied or claimed by another country. The act authorized the president of the United States to use military force to protect an island so claimed. Thus, in declaring himself sole owner of the guano islands, Norton was showing considerable business savvy. But it does not appear that he ever mined the islands for their natural resource.

As Emperor Norton I, he issued proclamations using the imperial "We" and printed his own money, which some sympathetic local eateries were said to have honored. Among his proclamations was one dated 1869 in which he ordered and directed "that a suspension bridge be constructed from . . . Oakland Point to Yerba Buena, from thence to the mountain range of Saucilleto." At the time, San Francisco was known as Yerba Buena, a name that has since come to designate the island in the middle of the bay, so it appears that Norton was proposing a bridge project that would have been a combination of today's San Francisco–Oakland Bay and Golden Gate bridges. Since he is not known to have had any special technical expertise, it is likely that Norton's proclamation repeated, and perhaps expanded upon, proposals put forth by contemporary engineers for suspension bridges in the Bay Area. Still, such proposals, requiring spans of the order of a mile in length, were really quite ahead of their time and beyond the state of the art of the late nineteenth century.

By the early twentieth century, suspension bridges spanning 1,600 feet were becoming common, and even longer-spanning cantilever bridges— those that exploit a bracket rather than a hanging principle of support— had been built or were under construction. In fact, in the early 1920s, when tens of millions of commuters a year were crossing the boat-choked bay by ferry, a plan to bridge the Golden Gate with a structure spanning 4,200 feet was moving forward, and so many proposals had been put forth to cross the bay between San Francisco and Oakland that a commit- tee was appointed to recommend the one best way. The scheme chosen went from the preexisting earthenwork pier known as the Oakland Mole to Goat (now Yerba Buena) Island—through whose towering rock a world-record-setting tunnel would eventually be bored—and thence followed the route of an underwater ridge of bedrock on to San Francisco.

Construction on both the Golden Gate and Bay bridges began in 1933—the height of the Great Depression—and for more than three years work on them was under way simultaneously. The independent Golden Gate Bridge and Highway District had been created especially to finance and build that crossing, but the Bay Bridge was essentially a state highway job, albeit the single most expensive public works project of any kind undertaken in America until that time. California state highway engineer Charles H. Purcell was appointed chief engineer of the project, but the bridge was effectively designed by committee—a view supported by the structure's lack of a unifying aesthetic between its east and west sections. There was a Board of Consulting Architects, chaired by Californian Timothy Pflueger. However, most of its recommendations were opposed by the bridge's design engineer, Glenn B. Woodruff. Even the color of the bridge became a point of disagreement. Civic groups objected to the engineers' initial preference for black, which Purcell agreed would highlight the heaviness of the structure. Alternative choices were gray and aluminum. The architects preferred the former, thinking the latter "thin and sheet-metal-like in appearance." The color used, aluminum gray, represented a compromise. While the architects did prevail in having the suspension-bridge tower design include bold diagonal bracing, they did not with their idea of decorating the massive common anchorage of the tandem structures with an art deco Egyptian figure that would have appeared to be holding back the pull of the suspension cables. At one point Pflueger asked, "Isn't there anything to this bridge but cheapness?"

Thanks in part to the restraint of the engineers, the Bay Bridge was completed ahead of schedule and under budget. Its final cost was $77.6 million, compared to the Golden Gate's $33 million. Remarkably by today's expectations, it was the Bay Bridge, arguably the more challenging engineering project, that was completed first, in October 1936; the completion of the Golden Gate followed in April 1937. In 1939, the Golden Gate International Exposition—that year's West Coast "world's fair"—opened on the purposely made Treasure Island, located just north of Yerba Buena Island, to celebrate the completion of "the world's two greatest bridges"—the Golden Gate and the Bay—certainly a defensible assertion by contemporary standards. Today the Golden Gate Bridge is universally known and recognized as an iconic symbol of San Francisco. However, at the groundbreaking for the Bay Bridge, former President Herbert Hoover called it "the greatest bridge yet erected by the human race."

But the Golden Gate Bridge, at 4,200 feet between its towers, was the longest single span of any kind in world, even though the Bay Bridge's total length of more than eight miles (about half of which is over land) made it the longest overall. One booklet published in the year of its

This postcard image shows the San Francisco–Oakland Bay Bridge in 1936 before it was opened to traffic. The Loma Prieta earthquake of 1989 revealed the structural vulnerability of the East Bay spans, which have been replaced with a modern viaduct and a unique self-anchored suspension-bridge span. After much controversy, many delays, and growing cost overruns, the new crossing was completed in 2013.

completion exclaimed that "the San Francisco–Oakland Bay Bridge is predicted to stand for a thousand years as the longest and largest bridge in the world!" Within two decades, the Lake Pontchartrain Causeway, near New Orleans, was reaching twenty-four miles over water, but by means of a series of many modest spans, the longest being only 180 feet in length. The Golden Gate would hold its longest-span record until 1964, when the Verrazano-Narrows Bridge, with a sixty-foot-longer main span, opened across the entrance to Upper New York Bay.

In the mid-1950s, when members of the American Society of Civil Engineers were polled to determine the seven modern civil engineering wonders of the United States, the Bay Bridge was named one of them. In 1994, however, when a similar poll was conducted worldwide, it was the Golden Gate, "one of the world's most revered and photographed bridges," that was listed among the Seven Wonders of the Modern World. Still, one distinction that the Bay Bridge does maintain is the volume of traffic it carries: at last count, an average of 280,000 vehicles daily (compared to just over 100,000 for the Golden Gate). That makes it comparable in traffic volume with the George Washington Bridge, which carries Interstate 95 over the Hudson River between New York and New Jersey. The distinction of being so heavily trafficked is among the principal factors that complicated the Bay Bridge's recent history and future form.

In the meantime, Emperor Norton had not been forgotten. The Emperor Norton Inn, a twelve-room residential hotel, is located within a couple blocks of Union Square; the San Francisco Brewing Company in North Beach produces Emperor Norton lager; and until recently the ice cream parlor at Ghirardelli Square offered an Emperor Norton sundae. A movement to rename the Bay Bridge for Norton, whose proclamations not only proposed bridges but also defended minorities and championed civil rights, fell on sympathetic ears in a city known for those values. According to city supervisor Aaron Peskin, "Emperor Norton was a model San Franciscan, extolling the virtues of tolerance, compassion and embracing diversity in our community." In 2004, to much derision by less appreciative citizens, the San Francisco Board of Supervisors voted—whimsically, it was later said—to rename the bridge after Norton. If it had been approved by the mayor, the proposal would have been forwarded to the Oakland City Council and then on to the California Legislature for their concurrence. What the legislature might have done would have been anyone's guess, given the animosity that had developed between Southern California and Bay Area politicians over what to do about the East Bay crossing.

THE RIFT BETWEEN THE Bay Area and the rest of California was not opened up by the Loma Prieta earthquake, but it grew wider when that event revealed weaknesses in the Bay Bridge. After the 1989 temblor, the vulnerability of the bridge, which is located within ten miles of both the San Andreas Fault and the Hayward Fault, could no longer be talked about in the context of geological and engineering theory alone. Something definitive clearly had to be done to protect the structure against the future "big one." A preliminary study done in 1992 had estimated that as much as $200 million was required for a seismic retrofit and in excess of $1 billion for a complete replacement. The 1994 Northridge earthquake, which did extensive damage to highway infrastructure in Southern California, prompted Caltrans to revisit its retrofit strategy for the East Bay Bridge. Subsequently, the transportation agency undertook a more comprehensive study that took years to complete, concluding that reinforcing the eastern portion would be so expensive (the better part of a billion dollars) and so disruptive to traffic that it made more sense to build from scratch a replacement. In the wake of the Northridge event, Caltrans' Seismic Advisory Board formally recommended replacement rather than repair. In early 1996, Caltrans announced that a new East Bay span could cost as much as $1.3 billion and that it would take another year to fully evaluate a retrofit alternative. By the end of the year, with estimates for retrofitting and rebuilding remaining essentially the same, the prospect of a replacement span grew more likely, and in early 1997 the decision was made to proceed with the design of a new bridge. Thus began an infrastructure saga of monumental proportions.

The existing East Bay crossing was undistinguished except for the steel cantilever that towered over the shipping channel just east of Yerba Buena Island. When first built, the cantilever's 1,400-foot span made it the longest in the United States, the second longest in North America, and the third longest in the world. As long as it stood, it remained among the top ten cantilevers of all time. The rest of the original East Bay crossing consisted of a series of relatively modest truss and girder spans connecting the cantilever to Oakland. By contrast, the West Bay crossing, which was to be retrofitted but not replaced, contains the dramatic pair of suspension bridges in tandem that share a common center anchorage, which—because it sits on bedrock 220 feet beneath the water at low tide—was in itself a most remarkable engineering feat. When opened to traffic, the 2,310-foot-span twin structures were the second-longest suspension spans in the world. (Only the 3,500-foot main span of the George Washington Bridge was longer, since the Golden Gate had yet to be completed.)

Initial design ideas for a new East Bay crossing did not do much to change perceptions of the relative aesthetic importance of the east and west parts of the bridge. When a proposed new viaduct design failed to gain broad popular support, the idea for a signature span—one that would be eminently distinctive and so raise the status of the East Bay crossing (and the city of Oakland)—was put on the table. A unique structure is incompatible with a quick, relatively low-budget solution, however, and it usually engenders debate over technical, aesthetic, and financial matters. As the argument over the bridge's appearance escalated, so did the cost, with the estimate for one two-tower, cable-stayed signature span option reaching $1.7 billion. To sort through all the complexities, the area's Metropolitan Transportation Commission, which was overseeing the East Bay project, established an Engineering Design Advisory Panel whose members included engineers, architects, seismologists, and representatives of Bay Area counties that would be expected to foot the bill for the incremental cost of building a more-than-utilitarian bridge. The panel recommended that an open design competition be held to generate ideas.

The transportation department advised that the signature span should be either a cable-stayed bridge—one whose roadway is supported by long, straight cables connected directly to the structure's tall towers—or a self-anchored suspension bridge—one with swooping cables for which massive anchorages are unnecessary—and issued a contract to a joint-venture engineering team to establish which design would be more appropriate. The team consisted of T.Y. Lin International, a distinguished San Francisco-based firm, and Moffatt & Nichol, which is based in Long Beach, thus involving both local and downstate economies. As is often the case in large projects, specialty firms were also involved as subconsultants, and the New York-based Weidlinger Associates, which has considerable experience with suspension bridges, became the team's structural engineer for that option. The charge from Caltrans was to develop for each bridge type preliminary designs to about one-third completion, so that they could be compared as to how they would look, how they would be constructed, how they would behave in an earthquake, and what they would cost.

In a self-anchored suspension bridge, the pull of the cables is opposed by the push of the roadway or deck, much the way the tension in a rubber band is resisted by the compression of the objects it binds together. Most self-anchored bridges look very much like a conventional suspension bridge, with its cables draped over two towers but without the massive anchorages that bookend it. The winning Bay Bridge design eliminated not only both of the anchorages but also one of the towers, resulting in a unique structure. Ironically, one of the five main options considered for the Goat Island-to-Oakland portion of the original Bay Bridge was a two-towered "self-anchoring suspension" span.

According to one of the principal designers at Weidlinger, the specification of a single tower was driven mainly by unfavorable underwater conditions, which would have required that a premium be paid for a second tower and its foundation. Among the other constraints was that no bridge tower could be higher than the 518-foot-tall ones of the West Bay spans. Also, unlike the existing crossing, in which eastbound traffic used the lower deck, the new bridge had to carry all traffic on a single level—a feature that obviated a repeat of the 1989 collapse of one deck onto the other. Finally, the cost of the signature structure was not to be any greater than that of the approach viaducts it would link up with. Since a cable-stayed bridge generally requires a taller tower than a suspension bridge of the same span, the latter type had an advantage. In the end, the self-anchored suspension bridge appears to have won out principally on aesthetic grounds. One of its principal designers described it as being "envisioned as a 'white line' across the Bay."

Not all structural critics agreed with either the motivation or the choice. Others believed that because San Francisco already had, in both the Golden Gate and the West Bay crossings, world-class suspension bridges, it was time that Oakland had one also. Christian Menn, the distinguished Swiss bridge designer responsible for Boston's signature Leonard P. Zakim Bunker Hill Memorial Bridge, termed the East Bay suspension-span solution an "architectural bridge." By this he meant a span whose design was driven by form, as opposed to an engineering bridge, the design of which derives from structural and constructional considerations. While admitting that "the appearance of the bridge is certainly very good," Menn presciently remarked that "the costs will be much higher than originally planned."

Caltrans officially chose the self-anchored suspension bridge design in late 1998, but debate would continue over details such as the inclusion of a bicycle lane, the structure's exact alignment, and access to Navy-controlled land on Yerba Buena Island to conduct geological testing. The transportation department's decision about the East Bay crossing was no doubt influenced by the distinctiveness of the design. By the end of the twentieth century, cable-stayed bridges had become commonplace, but self-anchored suspension bridges were rare. In fact, there were only about twenty in the whole world. (The first three in the United States were Pittsburgh's Sixth, Seventh, and Ninth Street bridges, known collectively as "the Three Sisters," built in the late 1920s.) Furthermore, the one proposed for the East Bay would be the longest of its type and the first asymmetrical one to be built in America. But with uniqueness also come uncertainties—of complications during design and construction and of cost, just as Menn had predicted. The initial estimated cost for the signature span alone was $740 million, which was almost as much as the $866 million that one Caltrans engineer estimated in 1997 that it would cost to replace the entire East Bay crossing with a viaduct. (At the time, the additional cost of demolition of the existing crossing was put at $40 million.) After further design work on both the self-anchored suspension and cable-stayed options, the overall cost estimate had reached $1.5 billion. Governor Pete Wilson wanted Bay Area residents to shoulder the bulk of the cost, with only $500 million coming from the state's transportation budget. The project was expected to be completed by 2004.

After the decision was made to go ahead with the self-anchored suspension bridge signature span, its full design could proceed, which usually results in changes being made as details are adjusted. Among the design changes that had to be implemented here was the incorporation of a bicycle path. Since no traffic lanes were to be sacrificed to do this, the bike lane was designed to be cantilevered off the south side of the deck structure. This destroyed the transverse symmetry of the deck and introduced a large eccentric load. To balance that load, a heavy counter-weight had to be incorporated along the north side of the deck. The overall design loads on the bridge structure were increased also by revised predictions of ground motion in the event of an earthquake. Since the deck of a self-anchored suspension bridge reacts against the cables, which it anchors, the greater the tensile forces in the cables, the greater the compressive forces in the deck. And the greater the forces along the deck, the more it must be stiffened lest it buckle. But additional stiffening means additional weight, which in turn increases the tension in the cables, which then increases the compressive forces in the deck, which consequently requires more stiffening. Breaking such a vicious circle naturally complicates the solution to the structural design problem—and increases the cost of design and construction.

In the meantime, construction on the much more conventional part of the crossing, which came to be known as the Skyway, was to proceed. At groundbreaking ceremonies, which took place in early 2002, Governor Gray Davis pointed out that the concrete viaduct, which was more than a mile long, was designed to survive an earthquake as great as the legendary one that hit San Francisco in 1906. The alignment of the new one-level dual viaduct paralleled the existing double-decker one, and so nearby construction cranes became a looming presence to travelers. The high cranes were especially visible from vehicles traveling on the upper (westbound) deck, and provided a strikingly regular arrangement when they were idle over holidays. The tower cranes, which were used to erect the piers and concrete segments of the Skyway, stood in formation and their booms looked not unlike the shouldered arms on an honor guard of Brobdingnagian soldiers rising out of the bay point-ing the way to the signature span's future location.

A call for bids to construct the unique signature structure went out in early 2003. By this time the estimated cost of the entire East Bay cross-ing had risen to about $2.5 billion, a level that, if known early on, would surely have favored just retrofitting the original bridge. However, things had gone too far to turn back, and so the term of a toll surcharge instituted to help pay for the project was extended. In spite of all the warnings, Caltrans continued to maintain, publicly at least, that bids for the signature span itself would come in close to the original estimate—that is, around $750 million. Unfortunately, only one bid was submitted—by a joint venture comprising American Bridge Company, based in Pennsylvania; Nippon Steel Bridge of Tokyo; and Fluor Enterprises, an engineering and construction company based in Aliso Viejo, California, that specializes in complex and challenging projects—and it was for a whopping $1.4 billion. Needless to say, there was considerable shock and subsequent speculation on what accounted for the difference. The rising cost of steel and concrete, due in part to the demand for those construction materials in China, was blamed, as was the sheer complexity of the project and the increased cost of risk insurance after September 11, 2001.

Whatever the reasons for the unexpectedly high bid, the cost of the overall project had now soared to more than $5 billion, which forced a reconsideration of the signature span. After months of speculation, the extraordinary bid was rejected on the day that it was to expire, September 30, 2004. Governor Arnold Schwarzenegger, who had defeated Gray Davis in the previous year's gubernatorial recall election, announced in December 2004 that he preferred to scrap the self-anchored steel suspension span and substitute an extension of the concrete Skyway. This would enable the entire project to be completed by 2012, the date at which the signature span was to be finished, according to supporters of the governor's position. But others warned that switching to a viaduct design at this stage would entail additional design and environmental reviews, which would threaten to delay the project further. Opponents believed that the governor's move was intended to force the Bay Area politicians to increase their region's portion of the project's funding if they wanted to maintain the signature span, a view supported by a Schwarzenegger administration spokesman's reference to Bay Area voters that "if they want a signature bridge, they are going to have to pay for it." A suggestion that was a compromise of sorts—namely, to change from a self-anchored suspension bridge to a more conventional cable-stayed bridge, which had already been partially designed and which could use the foundation that was already under construction—was not taken by the state government.

The repeated escalation of costs and slippage of the completion date were, needless to say, an embarrassment for Caltrans. The embarrassment became especially acute in the closing days of 2004, when it was reported that an audit ordered by the state legislature found that the transportation department had "concealed cost overruns . . . mismanaged the project and consistently underestimated expenses." At about the same time, news of the opening of a new bridge in France begged for comparison. The Millau Viaduct in southern France is a strikingly beautiful and graceful cable-stayed structure that closed the last gap in the superhighway between Paris and Barcelona. The overall concept should be credited to Michel Virlogeux, the engineer who designed the cable-stayed Pont de Normandie, which crosses the mouth of the Seine. However, news reports of the Millau Viaduct's completion attributed its design to the team's more visible and more widely known participant, the architect Sir Norman Foster. The 1.6-mile-long structure, whose tallest pylon is higher than the Eiffel Tower, was said to appear to float above the mist and clouds that often fill the Tarn river valley. Just as spectacular was the fact that the bridge was built in less than three years at a cost of $522 million, all financed privately. Investors expected to recoup their investment—and more—out of income from tolls. In contrast to the Millau structure, which proud locals were calling "the most beautiful bridge in the world," the East Bay crossing was threatening to be an undistinguished low-level viaduct that would, according to

The Millau Viaduct, which is more than eight thousand feet long and eleven hundred feet high, crosses the Tarn river valley in southern France. This stunning multi-span bridge represents what can be achieved in a successful collaboration between an engineer and an architect.

contemporary estimates, take three times as long to build and cost almost ten times as much as the French gem and was being called "the most expensive bridge in the world."

In spite of this, Bay Area politicians persisted in pushing for the self-anchored suspension bridge signature design for the East Bay crossing, and the passage of legislation increasing tolls on area bridges to pay for it sealed the deal. Construction continued on the viaduct leading up to where it would join the signature span. In late 2006, the last of the 452 concrete segments that went into making up the viaduct was hoisted into place, thereby completing the Skyway. With all its structural pieces in place, they could be secured by means of internal cables that give the multi-segmented viaduct the integrity of a single structure. The roadway surface could be laid down, railings and lights installed, lane markings and traffic signs put in place, and a major section of a modern bridge completed—at an estimated cost for this part alone of $1.3 billion. Of course, the Skyway would abruptly end at the gap left for the self-anchored suspension bridge, which remained to be built.

To avoid another embarrassment, the signature project was broken up into several subprojects in the hopes of attracting more bidders on a second try. The lower of two bids for the self-anchored suspension bridge was submitted by the joint venture of American Bridge Company, which participated in the construction of the original bridge, and Fluor Enterprises. The team's winning bid of $1.43 billion was just $20 million under the new Caltrans estimate of $1.45 billion. As accurate as that estimate was, it was difficult to find a candid updating of the mid-2005 cost estimate for the entire project, which then stood at $6.2 billion. And this figure was exclusive of interest and financing charges, which when added to the direct construction costs could bring the final total to around $12 billion.

In 2009, the old bridge was closed to traffic for the five-day Labor Day weekend, during which time it was connected to a detour viaduct previously constructed on Yerba Buena Island. Once traffic was rerouted to the detour, the cantilever section of the old bridge could be moved aside, making way for construction of the suspension span to begin. Although that work was still ahead and the completion date of the

bridge stood at 2013, in the spring of 2009 a group of architects attend-
ing an American Institute of Architects convention in San Francisco had
toured the construction site—before the old cantilever was removed—
and proclaimed the still-incomplete signature suspension structure and
its approach viaduct a new landmark for the city. Engineers tended to
reserve their judgment until the structure was completed, which finally
did occur in September 2013.

Months after the opening of the new span to traffic—its ten lanes
were capable of carrying 300,000 vehicles per day—criticism of the
project continued and remained in the news. A pair of reports prepared
for an investigative committee of the state senate reiterated that the new
bridge was safe. However, one of the reports, which had been prepared
by a panel of independent engineers after reviewing the design and
construction aspects of the project, warned that the collection of problems that had come to light could reduce the completed structure's

*The signature self-anchored suspension bridge in the foreground of
this aerial view forms the centerpiece of the new East Bay portion
of the San Francisco–Oakland Bay Bridge. In the background is
the old bridge, the center section of its cantilever span already in the
process of being dismantled.*

future reliability. Among their expectations was that "portions of the new bridge will likely require retrofitting throughout the life of the span." They also recommended a "robust inspection and maintenance program" that would increase maintenance costs over the structure's lifetime, thereby adding further to budget overruns that stood at $5 billion about a year after the new span opened, ten years behind schedule.

The other document, prepared by a veteran Bay-area investigative reporter, focused on problems of oversight, transparency, intimidation, and retaliation on the project. Engineers had reported being reassigned after raising questions about welding done in China, and others said they were "gagged and banished" for raising similar concerns over quality control. Bridge project managers were said to have told critics not to put in writing negative observations about the work. Such accusations were what had already led state politicians to call for a criminal investigation of Caltrans by the California Highway Patrol. According to one report, Caltrans was under investigation for "knowingly accepting substandard work at taxpayer expense, and retaliating against those who sought to bring problems to light." Infrastructure is expensive enough without being made more expensive by poor management decisions.

Among the more embarrassing developments in 2013 was the failed final tightening procedure of about a third of the long threaded steel rods designed to attach seismic shock absorbers to the bridge. This happened after the rods had been left exposed to the elements for five years. An alternative way had to be devised to take the loads that the broken rods could not, and this resulted in the design of saddle-like devices that detract from the graceful lines of the design. The incident also raised questions about the strength of more than two thousand other bolts and rods used in the bridge. Bolts were again in the news the following year, when it was reported that virtually all of the four-hundred-plus bolts anchoring the signature bridge's tower to its base were submerged in water, a situation not called for in the design. The situation was blamed on insufficient grout having been used. The chairman of the Metropolitan Transportation Commission, who described himself as a "problem solver," remarked upon learning of the submerged bolts, "I certainly wish we would stop finding problems to solve."

But new problems continued to be reported. In early 2015 SFGate, the sister website of the *San Francisco Chronicle*, described how, when the 525-foot-tall tower was first installed, its top was found to be eighteen inches closer than its bottom to the East Bay shore. To correct the leaning, high-strength-steel cables anchored in Yerba Buena Island were attached to the top of the tower and tightened to a high tension. The idea was to stretch over the course of a year or so the steel rods securing the tower to its base and so correct the problem. It was not unlike using a claw hammer to pull a nail just a little bit out of a block of wood—only on a very large scale. Unfortunately, as was discovered when one of them was later removed for inspection, the rods responded not just by stretching but by developing microscopic cracks. Since the extent of cracking in the uninspected rods remained unknown, this made predicting the tower's response to a future earthquake uncertain, which essentially nullified the whole point of building the new bridge in the first place.

Even after the new east span was opened to traffic, the old one needed to be disassembled. In the meantime, the 1997 $40 million estimate for demolition of the old bridge had risen to more than $270 million, and cost-saving measures were sought. One proposal involved leaving some of the old span's piers in place, perhaps to serve as supports for a recreational pedestrian bridge reaching out into the bay. Another justification for leaving the piers standing was so they might serve as a memorial of sorts to the old bridge, something that had been done in the late nineteenth century near Dundee, Scotland, where in 1879 the original Tay Bridge had collapsed under a railroad train crossing it in a storm. The stumps of the piers of the failed bridge stand to this day beside the replacement bridge.

In spite of the controversies, delays, scandals, and cost overruns associated with the design and construction of the East Span of the San Francisco–Oakland Bay Bridge, it was named one of five finalists for the 2015 Outstanding Civil Engineering Achievement Awards of the American Society of Civil Engineers. The bridge was recognized in large part for the distinction of its signature span, which at 2,047 feet made it the longest self-anchored suspension bridge in the world, and at 258 feet across the widest bridge of any kind anywhere. According to the ASCE in its announcement of the award finalists, "The $6.4 billion project has become the State of California's largest public works project to date." That was a fitting distinction to a crossing that, in its original configuration, was the nation's largest highway project. However, in contrast to that model project, the schedule slippage and cost escalation of the new one gave it dubious distinctions, which may well have kept it from winning the ASCE award. That distinction went to an obscure project very far from a bridge in San Francisco Bay: a $43 million state-of-the-art British research station located on an ice shelf afloat in the Weddell Sea, which is in Antarctica.

In the Undergrowth

ELEVATEDS, SUBWAYS, AND ENGINEERS

No MATTER HOW WELL designed, constructed, and maintained the roads, streets, and bridges of a large city or metropolitan area may be, the sheer volume of vehicles that they encourage and then must handle—especially during rush hour—for all practical purposes renders new infrastructure inadequate almost as soon as it is opened to traffic. In most long-established and densely developed cities, where downtowns were laid out well before the appearance of the automobile, there simply is no room to widen old streets or add new ones without major physical, commercial, and political disruption. This is why elevated highways began to be constructed: as a natural solution to a difficult problem. Many of them, however, turned into liabilities and infrastructural nightmares.

The idea of elevated roads and pathways is in fact an old one. Some walled cities, like Chester, England, retain to this day quiet walkways above the fray of traffic on the ground level below. The elevated location of the walkway on the Brooklyn Bridge puts pedestrians above the vehicular traffic, thereby giving them a more relaxing and less visually obstructed view than from other bridges, where people and vehicles cross on the same level. Indeed, it was the placement of the walkway on the Brooklyn Bridge that made walking across it one of the greatest pedestrian experiences in the world. As populations increased, separating people and vehicles continued to be a goal of engineers and planners.

In 1900, when bicycles were more common than automobiles, Pasadena, California, opened along a route that is now part of the Arroyo Seco Parkway, a mile-long elevated cycleway that was projected to reach all the way to Los Angeles. The rapid rise of the automobile caused the project to be abandoned, however, and soon there was talk of keeping motor vehicles and pedestrians on dedicated levels. An article in a 1913 issue of *Cassier's Engineering Monthly* presented "an effective and practical plan for handling the congested traffic in great cities." The scheme consisted of a "five-storied street" in which several levels of pedestrian arcades, walkways, and bridges between the upper stories of tall buildings looked down on the unimpeded vehicle traffic in the streets below.

As early as the 1930s, Oakland, California, had an elevated highway serving as an approach road to the San Francisco–Oakland Bay Bridge. Two decades after the bridge opened, the single-level elevated roadway was replaced with a double-decked one known as the Cypress Street Viaduct, designed to take more traffic off local streets. The reinforced-concrete structure collapsed in the 1989 Loma Prieta earthquake, claiming the lives of forty-two people. Not surprisingly, when the highway system was rebuilt, the design reverted to a single- rather than a double-deck viaduct.

Another early example of an elevated roadway was New York City's West Side Highway, which was built in stages between 1929 and 1951. It was closed in 1973 when a car and truck fell through the roadway and made it difficult to ignore any longer the raised road's long-neglected and deteriorating condition. Talk of relocating it fully underground had been going on for decades, but that did not come to pass; the highway was not totally dismantled until 1989. Even as New York had been talking about replacing its elevated highway, Boston was talking about erecting one of its own. The first section of the Central Artery opened in 1954, and the whole thing, completed in 1959, ultimately proved to be ugly to look at, unappealing to travel under, and frustrating to drive upon. Within three decades plans were being made to relocate it underground in a project that came infamously and derisively known as "the Big Dig."

In large cities, traffic and congestion tend to go together. This was true in the horse-and-wagon era and the situation has persisted from the early days of the automobile. The presence of pedestrians further exacerbated the problem. Among proposed solutions to traffic congestion was this "five-storied street," illustrated in a 1913 issue of the magazine Cassier's Engineering Monthly.

Back on the other side of the continent, San Francisco had long looked to an express connection between the Bay and Golden Gate bridges. It sounded like a good idea until a section of the Embarcadero Freeway, a double-decked elevated highway, was constructed right in front of the city's famed Ferry Building, obscuring it and providing all the evidence needed to prove it was a bad idea. With citizens and city leaders alike opposed to the continued development of freeways throughout San Francisco, some $280 million of federal aid was effectively turned down in favor of maintaining the bulk of the city as it was before the age of the interstate. The elevated Embarcadero Freeway was finally dismantled after being damaged by the Loma Prieta earthquake.

In Seattle, a double-decked section of Washington State Route 99 known as the Alaskan Way Viaduct had run through downtown beside historic Elliott Bay since the early 1950s. The structure was damaged in a 2001 earthquake and began to be demolished ten years later, to be replaced by an enormous tunnel beneath the downtown area connecting with a more robust viaduct south of downtown. The tunnel was to be bored by a 57-foot-diameter machine named Bertha, which was built in Japan and was the largest of its kind when it began operating in mid-2013. However, by the end of that year Bertha had become immobilized after encountering a buried eight-inch-diameter steel pipe that had not been anticipated by the machine operator. After resuming work for only a couple of days, the tunnel borer was stopped again when it overheated due to damaged seals around the main bearing. In order to gain access to the disabled machine and repair it, a large open pit was excavated, from which groundwater was pumped out to maintain a stable pit bottom. This procedure inadvertently drained groundwater from the surrounding area, disturbing a portion of the viaduct still in use and causing the settlement of about thirty historic buildings in downtown Seattle.

The troublesome pipe had been installed a decade earlier by the state—ironically, to monitor groundwater at the location. However, the obstacle was not mentioned in contract documents for the tunnel project, and so the company building the tunnel said it knew nothing of the buried pipe. From the construction company's perspective, this made the state responsible for the cost of repairs to Bertha and for the delay of the project. The state, on the other hand, was refusing to pay the anticipated repair bill because it claimed that the existence of the pipe was knowable through prior reports and studies. Such disputes are common in infrastructure work and in construction projects generally.

When the business end of Bertha was finally removed in May 2015 and the damage evaluated, more than seals were found to need attention. Broken steel casings and gear teeth meant that repairs would take longer than previously thought. The unprecedented size of the machine had contributed to its problems, and some of its inherent design weaknesses had to be corrected before it could be reassembled and put back to work. The Japanese builder of the machine assumed responsibility for fixing the flawed front end. Needless to say, the schedule slipped further, and completion of the tunnel was now put at 2017.

IT IS NOT EASY to imagine exactly what politicians, engineers, city planners, and just plain citizens of the early decades of the automobile age were thinking when they conceived of and approved of the elevated highways that became so uniformly disapproved of by the end of the twentieth century. No doubt the promise of immediately lifting at least some of the congesting traffic up off the streets and into the air and gaining structural efficiency had a lot to do with the thinking, but the longer-term effects that might have been foreseen were evidently not. This may have been due in part to positive initial experiences with separating local and through traffic, but the explosive growth of motor vehicles after World War II was not anticipated, at least to the extent that it was to occur and choke even elevated highways. It was then that the ugly underbelly of the elevated motorway dripping oil, rust, and concrete dust truly revealed itself.

Nevertheless, there were well-intentioned visionaries, among them Norman Bel Geddes, who was born in 1893 about forty miles west of Detroit, center of the nascent automobile industry. The unusual surname—Bel Geddes—was devised when Norman Melancton Geddes and Helen Belle Schneider married, and the portmanteau surname symbolized the joining of two creative minds. From an early age, when his mother took him to the opera, Bel Geddes had been fascinated by the stage, and his early success as a set designer led to his career in theatrical work. Successively, the man whom the *New York Times* once described as "the Leonardo da Vinci of the 20th century" became an industrial designer, a popularizer of streamlining, and the creator of the 1939 New York World's Fair General Motors "Highways and Horizons" exhibit known as "Futurama," which he described as "a large-scale model representing almost every type of terrain in America and illustrating how a motorway system may be laid down over the entire country." Futurama was viewed by five million fairgoers, making it then "the most popular show of any Fair in history." Visitors looked down upon the future from comfortable chairs that moved around the periphery of the model, which presented what the world of motorways might look like twenty years hence. Among the visions on display was an automated highway system in which automobiles essentially drove themselves, something that is being realized today in the form of Google and other autonomous vehicles. Bel Geddes attributed the success of the exhibit to the fact that "the people who stood in line ride in motor cars and therefore are harassed by the daily task" of driving, and "Futurama gave them a dramatic and graphic solution to a problem which they all faced." (The exhibit was so successful that General Motors produced an updated "Futurama II" for the 1964 New York World's Fair.)

Bel Geddes' post-fair book, *Magic Motorways*, and the 1940 General Motors film *To New Horizons* contained numerous sanitized illustrations of not necessarily original concepts of what the future could look like. Prominent among them was a view of a "street intersection—city of tomorrow," in which traffic appeared to be moving smoothly on one-way streets that were free of pedestrians, who were confined to elevated walkways or arcades highly reminiscent of those in the five-storied street concept proposed decades earlier in *Cassier's Monthly*. Bel Geddes' book contained several versions of the elevated roadway concept, including one that added a raised deck above the crest of a power dam that already carried traffic across a river, as well as a double-deck bridge

across a lake. Ironically, for all the futuristic intersections and magical
motorways illustrated in Bel Geddes' book, which gave no hint of the
coming opposition, the vast majority of depictions of vehicles were of a
distinctly late 1930s vintage. It was as if Bel Geddes had his eyes so
intently focused on the road that, in spite of his teardrop-shaped streamlining concepts, he did not think much about how the car in which he
was riding might look in 1960. Or perhaps that was to be the purview of
General Motors designers. After all, GM might not have wanted to
reveal to the public or to the competition what it imagined automobiles
twenty years hence would look like, would it?

As THE EARLY ELEVATED highways show, Bel Geddes and others were
not the first persons to think about separating pedestrians and vehicles,
and the concept predated even the elevated motor highway. Horse-
drawn wagons and carriages had been numerous enough in large cities
to congest the streets and make crossing them by people on foot problematic because of the traffic and also what horses leave behind. Mass
transit, in the form of horse-drawn omnibuses, horse-drawn rail cars,
and later electrified trolleys, promised some relief by multiplying the
number of people in vehicles and thereby keeping them off the street, at
least temporarily. But mass transit vehicles themselves contributed to
the ground-level congestion. Thus, some of the earliest light-rail systems
became elevated.

Elevated railroads to serve growing numbers of commuters were
chartered in New York City in the 1870s and 1880s. These "elevateds,"
"els," or "Ls," as they came variously to be known, provided some relief
to the volume of surface traffic, but the structural columns supporting
the tracks, stations, and ancillary equipment were often planted directly
in the street and so presented obstructions to the free flow of traffic
there. The convenience of the els for commuters came at considerable
cost in quality of life to those who lived beside the elevated structures,
which blocked out light, interfered with fresh airflow, carried deafen-
ingly loud trains right past apartment windows, and reduced property
values for the owners of the buildings.

In the mid-nineteenth century, underground transportation had been a much discussed idea in New York, as it was elsewhere. However, opposition to such a solution to the city's rapid transit needs arose on both technical and political grounds. Among the technical issues was concern over the polluting effects of smoke-belching steam locomotives in a confined space. Aboveground, the influential owners of the els and of street-level horse railways naturally opposed underground systems and had political clout. There were clearly opportunities for innovators who combined technical novelty and political savvy.

Alfred E. Beach, co-owner and editor of *Scientific American*, was a well-positioned inventor and journalist who in the late 1860s developed his idea to build a pneumatic subway powered by a large fan that drove a snug-fitting passenger car through a tunnel between stations as if it were a piston in a cylinder. He eluded political opposition by misrepresenting his plan to the state legislature, whose permission was required before construction could begin. Under the ruse of installing a system of message- and small-parcel-carrying pneumatic tubes beneath Broadway, he surreptitiously had a prototype of his subway constructed. When the one-block-long system (located near City Hall) was completed in early 1870, Beach unveiled his fait accompli to the public, who paid to travel in it as if it were an amusement ride.

In the meantime, the corruption of the New York political machine at Tammany Hall had been exposed, opening up the possibility that Beach's subway, even though illegally constructed, could receive a charter from the state legislature. But while waiting for that to happen, Beach continued to work on the intertwined technical and economic details of his system. The use of large fans and compressed air did not prove to be energy efficient, and Beach began to reconsider the use of steam locomotion underground, as had been done in London as early as 1863. But skepticism about the soundness of such an investment, combined with a poor economic climate, left Beach without sufficient capital to proceed. His subway was abandoned in 1874.

Nevertheless, New York's need for mass transit kept on growing. In spite of elevated railways being ugly, dirty, and noisy, their proven and

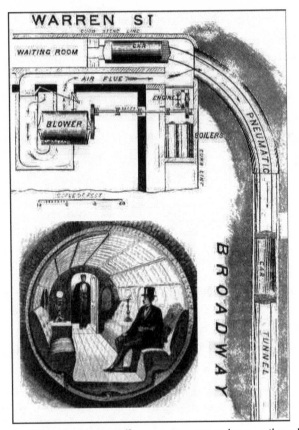

One means of alleviating traffic congestion was to locate railway lines
above or below street level. An early subway scheme was developed
surreptitiously into a prototype in Lower Manhattan. It was conceived
in the 1860s by Alfred E. Beach, an inventor and editor of Scientific
American. His scheme employed a tubular car that would fit snugly in
a circular tunnel and be propelled by compressed air.

less expensive technology kept them being proposed and built, and by 1880 els ran along Ninth, Second, and Third Avenues as far north as the Harlem River. The dream of a subway continued to be shared by many. One group of private investors established the Arcade Railway Company, which hired William B. Parsons as an engineer. But the company was not stable, and so Parsons and some of his colleagues left it to start their own New York District Railway. Both firms soon went bankrupt, leaving Parsons with a burning desire to develop a subway for New York, but with no organizational base. Parsons did, however, have family ties and civic connections.

William Barclay Parsons was born in 1859 into a socially prominent New York City family that traced its roots back to colonial times. After beginning his secondary education in Torquay, England, and completing it under private tutors while traveling about Europe, he returned to New York to attend Columbia College, from which he earned a bachelor of arts degree in 1879. He then studied at Columbia's School of Mines, from which he graduated in 1882 with the degree of civil engineer. He thus possessed the rudiments to design a significant part of the city's underground infrastructure, which remains in service today as a major component of the New York City subway system. However, before that would happen, his engineering talents would take him around the world to battle the forces of nature, but he would always return to New York. Indeed, he would be described by the *Times* as "city-born, city-bred, and city-minded," having through noblesse oblige exhibited the civic responsibility bestowed upon him by his ancestry and chosen profession.

Parsons used his reputation and connections to gain the ears of influential supporters for a subway. Among these was Abram S. Hewitt, who in 1886 defeated Theodore Roosevelt to become mayor of New York. Parsons also knew William Steinway, who had moved the family piano-making business across New York's East River and developed land around it in Astoria, Queens, running a horse railway to make the then-remote location more accessible to the city. Steinway's experience and interest in mass transportation made him a natural—even in light of a possible conflict of interest, which did not seem to matter much in the

The engineer William Barclay Parsons was responsible for designing and building the first New York City subway, which ran along Broadway for the length of Manhattan and eventually into the Bronx. He took on the Herculean task as a young man and later admitted that if he had realized how difficult the project would be, he might never have taken the job.

climate of that time—to be appointed in 1890 to New York's rapid transit commission by Mayor Hugh J. Grant, Hewitt's successor. Steinway was elected chair of the body, which came to be known as the Steinway Commission.

The chief engineer of the commission was William E. Worthen, a hydraulic and sanitary engineer who had served as vice president of the New York and New Haven Railroad and in 1887 was elected president of the American Society of Civil Engineers, which was then headquartered in New York. His career would not seem to have made him an obvious choice for an engineer of subways, but at the time there were few subways to serve as training grounds for specialist engineers. Experience in the design of water supply and sewer systems would have provided some useful background for underground work, but it may also have been Worthen's executive and professional-political achievements that had led to his appointment. Young Parsons, whom Mayor Hewitt had publicly declared to be a leading subway expert, became deputy chief engineer of the commission.

What underlay the island of Manhattan remained largely unexplored before the age of skyscrapers, which naturally required deep and firm foundations and thus deep and firm knowledge of the bedrock on which they would rest. In preparation for designing a subway system, Parsons had to map the ground through which its tunnels would pass. In the course of assessing the geology of New York, he had walked and plumbed much of the projected subway route and discovered that the island's bedrock—known as Manhattan schist—is not located at uniform depths underground. In Lower Manhattan and in what is now known as Midtown—and not by accident the two areas where the city's tallest buildings would become clustered—the rock is relatively close to the surface, but elsewhere on the island it is very deep. The undulating configuration and irregular composition of the bedrock would have presented technical difficulties to conventional tunneling, which led Parsons to opt for relatively shallow tunnels constructed by means of cut-and-cover technology, in which a large rectangular trench is excavated in the soil and then covered over at street level with a structural roof capable of supporting heavy traffic.

The recommendation stemmed from his direct geological exploration of Manhattan Island. This differed greatly from that of London,
which was underlaid with relatively soft clay through which deep
tunnels could easily be driven. But the deep-tunnel system demanded a
means to get people between street level and deep underground stations.
Elevators were necessary adjuncts that were costly to build and time-
consuming to use. Anyone who has experienced the London tube system
knows firsthand how frustrating it can be to wait for the large but
still limited-capacity lifts and how exhausting it is to use the stairs
when the elevators are out of service. Parsons' idea of placing New
York's subway close beneath the street surface enabled riders to move
quickly in and out of most of its stations by means of relatively short
flights of stairs. Designing infrastructure is nothing if not a complex
undertaking.

In 1891 the Steinway Commission announced its plans for the initial
New York subway routes, stretching from the southernmost part of
Manhattan to beyond the Harlem River to the north. The details of
construction were embodied in two engineering reports requested by
the commissioners, the reports differing in one important aspect. In
what came to be known as the Worthen plan, four tracks (two for local,
and two for express trains) were located on a single level, beneath
whatever gas, water, and electrical lines were buried in the street. The
Parsons plan proposed two distinct double-deck tunnels, one beneath
either side of the street, with the space between them reserved for the
pipes and conduits carrying utility services.

Four consulting engineers were engaged to review the two plans and
make recommendations. Although three of the consultants generally
preferred the Worthen scheme, the commission could not immediately
come together in endorsing a plan. Days dragged on into weeks. When
a call for bids to construct and operate a subway was finally issued,
there were no credible offers. Soon thereafter the commission abandoned all underground plans and sought to have an aboveground elevated railway constructed. But the Chamber of Commerce, representing
business interests, pushed for a subway subsidized by the government.
Former mayor Hewitt argued strongly against this proposal and called

for public ownership and private management. Out of such debates arose the Rapid Transit Act of 1894, which established a new Board of Rapid Transit Railroad Commissioners.

Among the board's first actions was to appoint Parsons chief engineer, a selection that was looked at askance because he was only thirty-five years old at the time. He would later look back on the appointment and say he was glad that he was not older, because he doubted that he would have accepted had he realized the magnitude of the task. Some of his friends even pitied him for chasing an unrealizable dream.

After his appointment, Parsons traveled to Europe to visit and study under- and aboveground railway systems in London (where electric power had been adopted only a few years earlier), Liverpool, Glasgow, and Paris. His subsequent report to the Steinway Commission contained several key features that he felt important for a subway in New York: the use of cut-and-cover rather than deep tunneling technology, the use of third-rail electric power rather than steam engines, and the use of a four-track express system rather than having all trains stop at all stations. These became key features of the New York subway system.

The engineering plan adopted in 1895 incorporated Parsons' fundamental design recommendations—with the notable addition of locating all four tracks on a single level—but the beginning of construction had to wait for a franchise to be awarded, and New York politics and lawsuits kept the process from moving forward. In the meantime, as new underground systems were opened in Boston and Budapest, Parsons grew uncertain and restless. In 1898, while the New York subway plan languished, he went to China to lay out an almost thousand-mile railroad through the relatively isolated Hunan province, where for many of the local people he was "the first foreigner ever seen." The Chinese project was financed by a syndicate that included J. P. Morgan, whom Parsons also knew socially. The engineer was in Canton in late 1899 when he learned that approval had finally been given to the New York subway project, and he went back home posthaste to build it.

In 1900 a construction contract was signed, and in 1902 the Interborough Rapid Transit Company was formed to operate the subway. Four and a half years later, the first ten miles of what came to be known as the IRT—running from City Hall to 145th Street in Manhattan—was opened to much fanfare, vindicating the disruptive cut-and-cover construction that had come to be known as "Parsons' ditch." Before long, the reach of the subway system would be expanded by the addition of the competing Brooklyn-Manhattan Transit Corporation (BMT) and the Independent Subway System (IND). By the 1940s, the New York City subway system consisted of more than seven hundred miles of track; in 2014, it accounted for 1.75 billion rides annually (5.6 million each workday) taking place among its more than four hundred stations.

AFTER HIS SUBWAY WORK, Parsons became involved with other transportation projects, including the Panama Canal and the Cape Cod Canal. In 1917, at the age of fifty-eight, he left the security of his domestic engineering practice to lead the U.S. Army's Eleventh Engineer Regiment building roads, bridges, docks, and railroads in France during World War I.

In his retirement, Parsons engaged in scholarly pursuits. In 1920, drawing on his military experience he published *The American Engineers in France*. Two years later he published *Robert Fulton and the Submarine*, reflecting his admiration for the steamboat pioneer and his achievements. Parsons' magnum opus, *Engineers and Engineering in the Renaissance*, appeared posthumously in 1939 and represented the fruits of a project to which he devoted much of the latter part of his life. Even in his historical writing, Parsons the engineer was ever present. In a chapter on the pre-Renaissance period, he proposed that the long preparation for the sudden rebirth had its origins in the fifth century with the fall of Rome, and he likened the process to tunneling. According to Parsons, the first half of the fifteenth century corresponded to the driving of tunnels and filling them with powder, with the final spark that set off the Renaissance being the capture of Constantinople by the Turks in 1453, a specific event that "swept away the remaining vestige of the Empire of the East, and left Europe face to face with a new order." As Parsons wrote:

The end of the Middle Ages and the beginning of the Renaissance came so suddenly that it is possible to fix a date, and the transition was so abrupt as to appear due to a single impulse. But this abruptness is fairly comparable to the culmination of the work of an engineer who, having laboriously excavated tunnels and galleries and filled them with high explosives, sends an electric spark to detonate the mass and thus in an instant of time blasts away the side of a mountain.

Parsons felt strongly that engineers were more than technical people with narrow specialties. In addition to their engineering expertise, they had to be versatile managers with well-rounded experience. Throughout his life Parsons strove to emphasize that being an engineer was not incompatible with being a learned gentleman. He was considered a Renaissance man, not only for his broad education, but also for his wide-ranging interests and achievements. Engineers are not as often memorialized the way politicians are, but Parsons Boulevard in Queens is named for the engineer and, beneath the street, it designates the antepenultimate station on the IND's F line, a fitting tribute to the Renaissance engineer responsible for establishing the standard for the New York subway system. His name also lives on in the well-established engineering firm Parsons Brinckerhoff, which is involved in infrastructure projects around the world.

No professional works in isolation, especially with respect to infrastructure projects. Matters of public health, property ownership, and technical problem solving come together as much in developing roads and subways as they do in providing ample clean water supplies. At the same time, not all engineers have been wholly likeable characters, having to work as they necessarily did in the real world where technology, finance, and politics intersect. They had to be prepared to negotiate and compromise, giving up a pursuit of the perfect for the better good. And they had to understand that the work of construction takes place in heat and cold, in rain and snow, in ice and mud. The making of infrastructure is neither a clean nor a comfortable enterprise. Nor is it safe and sure. It can be worse to watch than sausage making.

Nor is the planning and making of infrastructure necessarily even-handed. The urban historian Clay McShane has written that many of the engineering pioneers "treated pedestrians as second-class citizens." He characterizes Morris McClintock's 1925 work *Street Traffic Control* as the "first traffic engineering textbook" and says of its author that, "of all early traffic engineers, McClintock consistently expressed the most concern with safety," for pedestrians as well as for riders in vehicles. Nevertheless, even he saw people on foot, especially when jaywalking, as contributors to traffic congestion. At one point McClintock called crowds of pedestrians spilling over into the intersection while waiting for the light to change as "warring elements" against vehicles and thought that the people on foot should be physically separated from the traffic in the street. Elsewhere he advocated pedestrian bridges across especially busy streets and second-story arcades of the kind that Bel Geddes would later promote. Infrastructure—and its use—often returns to its beginnings.

8

Just as Fair

LOCATION, LOCATION, LOCATION—AND FINANCING

MONEY, POLITICS, AND INFRASTRUCTURE go together. Even when the money comes from private investors, politics is involved through concession, regulation, and legislation. Throughout most of twentieth-century America, roads were widened and new bridges built almost exclusively when politicians appropriated the funds to have the work done. Bridges, perhaps because of their largely self-contained nature, stand symbolically for infrastructure generally and often serve as surrogates for larger issues relating to it, and New York City and its environs present a concentrated example of how complex and interrelated those issues can be. The story of the replacement for the obsolete and deteriorating Tappan Zee Bridge located just north of New York City encapsulates much of what is involved in large infrastructure projects of all kinds and in all places. Like a good movie, the drama of the aging bridge across the Hudson River has elements of intrigue, suspense, and hidden agendas that, when exposed, reveal much about not only the nature of political power and influence but also often the futility of opposition to it.

The New York metropolitan area is an infrastructure miracle or nightmare, depending on your point of view. Its interconnected roads, bridges, and tunnels make it a vehicular broadband Internet, but one that can easily get overloaded and go down. It is full of traffic bottlenecks, and among the worst is the Tappan Zee Bridge, which carries the New York State Thruway—Interstate 87—across the Hudson River about twenty-five miles north of Upper New York Bay, into which the Hudson flows. Between the Tappan Zee and the bay there is only one other overwater crossing of the river proper, the George Washington Bridge connecting New Jersey and New York via Interstate 95. The Verrazano-Narrows Bridge, which takes Interstate 278 across the entrance to the upper bay known as the Narrows, carries traffic between the New York City boroughs of Staten Island and Brooklyn. Each of these last two crossings had the longest main span of any suspension bridge in the world when completed, the George Washington in 1931 and the Verrazano-Narrows in 1964. At the beginning of 2014, when the Verrazano was celebrating its fiftieth anniversary, the main spans of these bridges were still respectively the twenty-fourth and eleventh longest in the world. There are also two vehicular tunnels under the Hudson that connect New Jersey to Manhattan: the Holland Tunnel, which carries Interstate 78 into the city in the vicinity of Canal Street, and the Lincoln Tunnel, which comes in at Thirty-Ninth Street. These few fixed crossings of the river south of the Tappan Zee are critical to the road network that links New York City to points west.

Transportation problems are nothing new to the city that, until the last years of the nineteenth century, comprised only what are now known as the boroughs of Manhattan and the Bronx, separated by the relatively narrow and therefore relatively easily bridged Harlem River. But for a long time before that, the city's western boundary, the Hudson River, and its eastern boundary, the inaccurately named East River (which is really a tidal estuary connecting Upper New York Bay with Long Island Sound), were wider than bridge-building technology could span in a single leap, which was necessary because of the extensive boat and ship traffic around Manhattan Island. The East River, being about half the width of the Hudson, was the first to be attempted. Proposals were put forth in the early 1800s, and by midcentury there were numerous ideas on the table for connecting Manhattan to Brooklyn and Queens, both of which are part of Long Island. The dream became reality with the completion of the Brooklyn Bridge in 1883, making travel between New York and Brooklyn convenient and dependable. No longer did ice on the river or fog in the air hamper commuters by interrupting ferry service, so the two separate cities began to function more like a single one. In 1898, just fifteen years after the completion of the bridge, the City of Greater New York—consisting of the five boroughs of Manhattan, Brooklyn, Queens, the Bronx, and Staten Island—was formed, thus demonstrating the important role that infrastructure can play in shaping government and politics.

With the success of the Brooklyn Bridge—and the traffic it encouraged between Lower Manhattan and Brooklyn—there was soon a heightened desire and indeed a need for additional fixed crossings of the East River. The unification of the boroughs expedited the political process whereby bridges were planned and funded by the city, and so in the first decade of the twentieth century three new major spans were opened across the East River: the Williamsburg (1903) and Manhattan (1909) suspension bridges connecting Lower Manhattan and Brooklyn, and the Queensboro Bridge (1909), a cantilever connecting midtown Manhattan and Queens. Today, these centenarian overwater structures— along with the more recently built Triborough Bridge (1936), Queens-Midtown Tunnel (1940), and Brooklyn Battery Tunnel (1950), plus the Verrazano-Narrows Bridge—are among the principal vehicular traffic arteries joining the boroughs.

Although Manhattan and New Jersey had been joined by Pennsylvania Railroad tunnels by the end of the first decade of the twentieth century, the Hudson River remained without a bridge over it or a vehicular tunnel under it until the Holland Tunnel opened in 1927. The tunnel's being named in honor of its chief engineer, Clifford M. Holland, who died just before the first tunnel tube was holed through, makes it a rare example of a piece of infrastructure named after an engineer. Although a bridge would likely have had more traffic capacity than a tunnel, debates over the limits of bridge technology and the urgency of the situation argued for a tunnel to relieve the congestion of Hudson River ferries, which in the 1920s were carrying about 30 million passengers a year.

It had long been difficult to resolve jurisdictional disputes and coordinate infrastructure planning between New York and New Jersey, and so in 1921 the bistate Port of New York Authority was established, with each state's governor appointing equal numbers of members to the governing board of the port district. Many New Yorkers and New Jerseyites persisted in referring to the agency as the "Port of Authority," and this abusage long served as a kind of shibboleth for identifying natives of the area. More than this, New Jersey's desire for equal billing led to the 1972 name change to Port Authority of New York and New Jersey. It is responsible for the bulk of the region's transportation infrastructure, which includes tunnels, bridges, airports, and seaports within a twenty-five mile radius of the Statue of Liberty. The Port Authority planned and built the World Trade Center in Lower Manhattan, its originally planned location on the east side of the island having been moved to the west side so that New Jersey could more readily share in its economic benefits—a prime example of politics determining the location of infrastructure.

BY POLITICAL DESIGN, THE Tappan Zee Bridge was located just outside the twenty-five-mile radius of influence of the Port Authority. This explains in part why the bridge is located not at a narrow section of the Hudson but rather at one of its widest parts. The bridge, which carries the New York State Thruway between Rockland and Westchester Counties, is thus uncommonly long for a river crossing, stretching about three miles from end to end, with its longest span being just over 1,200 feet. Siting a bridge at a wide part of the river may be counterintuitive, but generally where a river is wide it is also relatively shallow. This can reduce the cost of constructing foundations, so many piers can be relatively close together and thus result in many modest spans. Drivers and passengers crossing the Tappan Zee Bridge are reminded of the crossing's many modest spans by the regular and frequent sound of tires rolling over road joints atop the piers. Unfortunately, poor riverbed conditions complicated foundation design, providing merely another instance where a political decision made technical accomplishments more difficult than they should have been.

The name Tappan Zee derives from the name of the Tappan Indian tribe that lived in the area and the Dutch word *zee*, meaning "sea," alluding to the expansive width of the river at the location. In 1950, the Port of New York Authority wished to build a new bridge across the

Since 1955, the Tappan Zee Bridge has carried the New York State Thruway across one of the widest reaches of the Hudson River. The crossing's aging steel cantilever had long been considered structurally deficient and functionally obsolete when a competition was held to design a replacement for the entire bridge.

Hudson River at Dobbs Ferry, New York, which is just inside the twenty-five-mile limit of its jurisdiction. Then New York governor Thomas E. Dewey effectively vetoed the proposal, and a plan to build a bridge at Tarrytown went forward. This location, being outside the Port Authority's jurisdiction, meant that toll revenue would go not to the bistate Port Authority but to the New York State Thruway Authority. The structural design called for a steel-cantilever main span flanked by viaduct-like approach spans. Among the innovative features of the bridge were its hollow concrete caisson foundations, whose buoyancy would help support the main span and thus reduce costs. Construction began in 1952 and the bridge was opened to traffic near the end of 1955.

Although commonly known by its original name, the Tappan Zee Bridge, since 1994 the official name of the crossing has been the Governor Malcolm Wilson Tappan Zee Bridge, even though Governor Wilson played no significant role in the bridge's history. Appending his name to the Tappan Zee was done to honor him politically on the occasion of the twentieth anniversary of his one-year term as governor. In fact, Wilson was never even elected governor: he rose from the position of lieutenant governor to serve out the remainder of Nelson Rockefeller's term when he resigned in 1973 to spend more time on national and international policy issues. Wilson, in an effort to continue in the governor's office, ran for election himself but lost. Perhaps there was among some state employees lingering resentment of his name being attached to the Tappan Zee structure, for when signs bearing the bridge's new name were unveiled on the South Nyack and Tarrytown approaches, it was evident that the governor's first name was misspelled as "Malcom." The signs had to be replaced at a cost of $3,000 each.

In the early years of the twenty-first century, the Tappan Zee, which has been called "one of the ugliest bridges in the East," had more serious problems than signs containing misspellings and general lack of aesthetic appeal. One infrastructure critic said that it was "being held together with glue and duct tape." It would certainly have taken more than those quick fixes to put the bridge back in mint condition. When it opened to traffic, on average about 18,000 vehicles a day used the crossing; approaching its sixtieth anniversary, it carried about 140,000 (and on peak occasions as many as 170,000) vehicles daily, which meant it was suffering the abuse of traffic volume on an order of magnitude greater than when the structure was new. At such traffic levels, it served about 50 million vehicles each year, which collectively paid about $650 million in tolls.

The traffic volume and intensity was taxing the structure's capacity, not only in number of vehicles but also in structural endurance. Every vehicle, and especially a heavy bus or truck, that crosses any bridge causes the structure's fabric to flex and creates a situation in which so-called fatigue cracks can be initiated and grow. A bridge is designed to last a certain number of years under such conditions, and the intended lifetime of a bridge such as the Tappan Zee when it was on the drawing board was about fifty years. As the bridge approached its golden anniversary in 2005, its owner, the New York State Thruway Authority, reportedly refuted that it was built to last only that long. But a few years later, perhaps because it could smell the federal cash that was likely becoming available for a replacement, the Thruway Authority began to speak of the bridge as "reaching the conclusion of its useful life." It was indeed technically considered functionally obsolete in that among other things there was nowhere for disabled vehicles to pull over and allow traffic to continue on its way. Still, the Thruway Authority's executive director insisted that the bridge was "perfectly safe" because inspection—intended to catch the development of corrosion and cracks and other problems before they become dangerous—was being done regularly. Furthermore, maintenance—designed to arrest the progression of any corrosion and cracks that are found—was keeping pace, even if the cumulative cost of it had long surpassed the original cost of the structure.

The Tappan Zee also fell into the category of bridges known as "fracture-critical," as it did not possess structural redundancy. If an essential part were to fail—whether because of an unstable fatigue crack or other weakness—there was not another part capable of taking up the slack, and so the entire bridge was likely to collapse. Some eighteen thousand such bridges exist in America. The Interstate 35W bridge over the Mississippi River in Minneapolis was a fracture-critical structure, and when it collapsed without warning under rush-hour traffic in 2007, the incident naturally led to increased scrutiny of all fracture-critical bridges, including the Tappan Zee. Calls were made for the design and construction of a replacement bridge over the Hudson at Tarrytown. However, advocates for a new bridge wanted more than just something that could carry more cars, busses, and trucks with fewer delays: they also wanted the new structure to accommodate mass transit in the form of a dedicated bus lane or a light-rail line. And, as with the East Bay span of the San Francisco–Oakland Bay Bridge, improving the aesthetics of the crossing was also a highly desirable if not essential objective. Even without the complications of such amenities, which could delay design and construction, the bridge was expected to cost at least $5 billion to build. With bus lanes and rail tracks, the price tag could reach $16 billion, according to a Federal Highway Administration estimate at the time.

In 2011, the case of the Tappan Zee Bridge came to the attention of the White House, and that October President Barack Obama announced that the proposed replacement Hudson River crossing would be among fourteen major infrastructure projects that would be "fast-tracked" to encourage development and promote job growth for the construction industry, which had been ailing since the housing bubble burst. A federal review of environmental impact would go forward on an accelerated schedule and the availability of federal funds for construction would be expedited. As planning and review progressed on a new Tappan Zee Bridge, which was to have at least eight traffic lanes compared to the existing structure's seven, newspaper reports began to quote officials that the price tag might rise to $6 billion. The situation reminded observers of the contemporaneously ongoing story of the East Bay span.

In times past, the next step in the process of acquiring a new bridge involved a design either by state highway engineers or a specialist consulting firm engaged by the state. When the design was complete, requests for proposals would be announced and published in trade journals of the construction industry. Construction companies interested in bidding on the project would submit their sealed bids by a stated deadline, and they would be ceremoniously opened and the job given to the lowest bidder unless other criteria had been stated also to be taken into consideration. That serial process not only took time but also had the potential of sometimes leading to bridges of unremarkable aesthetics, and at other times to protracted legal disputes between the designer and the builder over whether the plans or the construction was at fault when something did not match up or work properly.

To preclude such conflicts, a process known as "design-build" had increasingly been adopted in the United States: designers and builders form a team that bids on a project as an entity and thus has responsibility for both the design and the construction. Any conflicts that might arise will supposedly be resolved quickly and amicably within the team, for otherwise the partners share in any loss associated with delay or dispute. New York State passed a law in December 2011 to allow such contracting, and so building the new Tappan Zee, the state's "largest-ever bridge project," was to be among the first design-build projects to be overseen by New York state employees.

Only four qualifying partnerships and joint ventures received formal requests for proposals. The short-listed teams had about four months to prepare and submit their detailed plans and cost figures. According to one team member, "We've had more time on $50-million jobs," and yet this one had been estimated to be worth about a hundred times as much. The state agencies would have to review and evaluate all four proposals in a similarly short period of time if a contract were to be offered in time for work to begin before the November 2012 elections, as politicians wished. New York governor Andrew Cuomo—no doubt along with some politicians in Washington—was said to be especially keen to see that work was started by then.

Of course, it takes time and money to come up with a design and independent cost estimate and to prepare a proposal in response to any request. In the case of the Tappan Zee project, each qualified proposing team reportedly was to receive a stipend of $2.5 million for its efforts. Whether any such stipend ever covers the actual cost of preparing a proposal is a debated topic, and it surely depends upon how much free rein the bidders are allowed. In the case of the Tappan Zee Bridge Hudson River Crossing Project—to use its official name—among the constraints placed on the bidders by the state were that the bridge be located just a few hundred feet north of and beside the existing bridge and that it consist of two separate but equal structures, one carrying eastbound and the other westbound traffic. Each structure was to accommodate four lanes of traffic, have a left and right breakdown lane or shoulder, and have a lane dedicated to pedestrians and cyclists. (The significant additional expense of including light rail put it on the back burner.) It seemed likely that all four qualified teams would go forward and produce a document endeavoring to convince those reviewing it that it would give the state the best deal. However, only three of the teams met the July 27, 2012 bid deadline.

In addition to evaluating the foundation and structural design, construction plan, and projected life span of the bridge, selection criteria for choosing the winning team were to take into account future transportation options, environmental requirements, and how the design-build team planned to work with a broad range of stakeholders. The lowest of the three bids was not necessarily expected to win the contract, for in addition to bottom-line price each proposal was to be evaluated on the basis of the maintainability of the proposed bridge and on its "architectural iconism," which effectively allowed aesthetic judgment and current structural fashion to enter into the evaluation.

In the meantime, after failing earlier in the year to secure a federal loan for about $2 billion, New York State had reapplied to Washington, this time for a $2.9 billion loan. In mid-September 2012, Governor Cuomo announced the formation of a "selection review team," which included a sculptor and the director of the Metropolitan Museum of Art. This group was to evaluate the aesthetics of the proposed designs and assist a larger review team, which included technical experts and local community leaders, to choose the proposal that would give the "best value" and "ensure the new bridge is the best choice and fit for the region." The review team had the option of recommending to the governor that he accept one of the proposals as written, that he authorize negotiations with selected bidders, or that he initiate a call for "a best and final offer" from one or more of the bidders.

For at least a decade, and even as the proposals were being evaluated, other facets of a new bridge had come under scrutiny, including how its construction would affect the sturgeon that populate the river. Activists and public interest advocates complained that the request for proposals was issued before the deadline for public comment on the project, thus not taking fully into account alternatives to a new bridge and what ordinary people considered important. Among the things that continued to interest area residents was what would happen to the old bridge when the new one was in place. Many wanted to save the $150 million it could cost to demolish the old structure and use the money to convert it into a thirty-acre park that would stretch along the three miles of the crossing.

The three proposals for a new Tappan Zee crossing were kept secret while contract negotiations took place with the apparent front-runner. In early December 2012 the three designs and their costs were unveiled. It was little surprise to bridge watchers that all three proposals were for cable-stayed structures, in which the longest spans are supported by cables running directly from the towers to the roadway. The structural category allows for wide variations in tower geometry and cable arrangement. One design was described in the popular press as resembling "pairs of tuning forks," another as a "crowd of sail boats" in a regatta, and the other as a "cream-colored version" of the Golden Gate Bridge, even though that iconic span is a conventional suspension and not a cable-stayed bridge. In revealing the competitors, the advisory committee described how long each option would take to construct, which ranged from just over five years to almost six; the amount of dredging that would be required, from just under one million to just under two million cubic yards; and the price, which ranged from a low of just over $3 billion for the first of the options to about $4 billion for the other two. That all the bids came in under the $5 billion estimate reflected the depressed state of the construction industry. Companies were willing to risk a financial loss on a big job rather than be without work.

Just before the end of 2012, the Thruway Authority board of directors approved a $3.14 billion contract for the tuning fork design proposed by the team of Tappan Zee Constructors, a consortium headed by the international construction company Fluor. The design allows for the innermost two of the four outward-leaning pylons to be connected at their tops at some future time and thereby provide a support system for a third bridge to be constructed between the proposed two parallel ones. This third bridge could carry a pair of railroad tracks. But in the meantime engineers would have to deal with building the pair of highway bridges, which in 2014 was called "perhaps the largest and most challenging bridge project currently under construction in the U.S."

Construction management, which could add as much as $800 million to the cost of the job and thus bring the total to just about $4 billion, was to be handled by the HNTB Corporation. The state of New York had signed off on the contract early in 2013, and everything seemed in place except the financing. The federal loan was still pending, as was the issuance of bonds to be covered by toll revenue, but groundbreaking for the projected five-year project occurred without the money in place. The bulk of one low-interest loan expected to come from monies

The new Tappan Zee Bridge will feature a pair of identical side-by-side cable-stayed bridges that collectively provide a signature structure for the Hudson River crossing. At some future date, the inside pairs of canted pylons may be used to support a deck dedicated to carrying a light-rail system.

associated with the Clean Water State Revolving Fund was largely denied by the Environmental Protection Agency, with only $29 million of the $511 million asked for being approved. New York had proposed using $110 million of that amount for river dredging and $65 million to remove the old bridge when the new one was in service, but these were considered inappropriate uses of money intended for sewer and clean-water projects.

As of the fall of 2014, the financing was still not in place. However, when the state of New York received a windfall of $4.2 billion from a number of separate legal settlements with insurance companies and banks, the governor considered using some of it to pay for the bridge project. Whether he would be able to do so remained to be seen. In the meantime, construction continued.

On February 2, 2015, the White House's proposed federal budget for fiscal year 2016 was released by the Office of Management and Budget to the usual media hype. In Washington, D.C., news photographers caught the arrival of pallets of the printed document and its distribution to eagerly awaiting reporters, political staffers, lobbyists, and analysts. Earlier that same day, in Punxsutawney, Pennsylvania, the celebrity groundhog named Phil had come out of his burrow and seen his shadow, and so predicted six more weeks of winter, which was easy to believe amidst the series of snowstorms that had been battering the Northeast. What was less credible for some in Washington and elsewhere around the country was the $4 trillion federal budget outlined inside the heavy books.

However, in addition to the staggeringly large number, what caught the attention of many budget watchers and commentators was the photograph on the cover of the books of numbers. Whereas in years past federal budgets were issued without cover photos, this edition featured a bridge—the old Tappan Zee! In the photo, which appeared to have been taken years earlier and definitely not during rush hour, the bridge actually looked good: its lacy upper girders showing no obvious rust; its roadway pothole-free and uncluttered with traffic. What a nice-looking bridge! As effective as images of rusting and crumbling roads and bridges might be in arguing for more spending on them, such images would send the wrong signal on the cover of a budget promising to reclaim our infrastructure. The White House was evidently attempting to give a clear indication that infrastructure was an unthreatening and important component of the new budget, but a close look at the numbers raised the question of how important. The total proposed for transportation infrastructure projects was $478 billion, but over six years. That's about $80 billion a year, which is about 2 percent of the total annual budget. Forty percent of that amount was already coming from the existing federal gasoline tax, and the rest of it was to come from a new tax on foreign corporate earnings held overseas, a concept known as "repatriation." Whether Congress would go along with the idea remained to be seen.

Ironically, putting a not-unattractive image of the existing Tappan Zee Bridge on the cover of the federal budget was but a reminder to those who knew its story and its financial situation that the state of New York was crossing its fingers that the old bridge would last until

The president's budget request for the U.S. government for the fiscal year 2016 filled four paperback volumes. As shown in this detail of the cover of the document, it was illustrated with a photograph of the Tappan Zee Bridge, whose replacement the Obama administration had identified as a priority project. Of the almost $4 trillion total budget, $478 billion over six years was for transportation infrastructure.

the new one was completed and that the funding for the signature span would be forthcoming. At the time of release of the federal budget, only $1.6 billion in federal funds had been secured toward covering the $3.9 billion basic price tag of the new bridge. As much of a political backdrop that the old structure had provided for photo opportunities and now a budget cover, its financing remained to be fully arranged.

What everyone knew but few seemed to discuss was that the financing plan of last resort was to raise tolls when the new bridge opened.

The action would be justified as necessary to pay off bonds floated to raise cash for construction, and the promise would be that when the bonds were retired, the tolls would be rolled back. It was a familiar scenario, and one recalled bitterly on the occasion of the fiftieth anniversary of the Verrazano-Narrows Bridge. Old-time Staten Islanders swore that they were promised that once that bridge was paid for, passage across it would be toll-free for them. Unfortunately, bridges and many other pieces of infrastructure are never really paid for, in the sense that there are continuing maintenance costs to keep them in good structural health and proper operating order. Such mundane details are not what politicians emphasize—or even mention—when they are presenting a new project or speaking at a ribbon-cutting ceremony. Furthermore, as long as tolls are being collected, they provide a revenue stream that not only benefits the bridge but also financially props up other transit projects, no matter how indirectly related to the bridge. In the case of the Verrazano-Narrows, the approach of its golden anniversary had driven historians to the archives, but no one could find any official document specifying an end to toll collection, ever.

9

Perhaps the Better Claim

ICONIC, SIGNATURE, AND STUNNING SPANS

AMONG THE COMPLICATIONS FOR the builders of the Tappan Zee replacement bridge was how they would deal with the noise that would accompany construction. More than one thousand steel piles, some over two hundred feet long, were to be driven into the riverbed. Not only would the noise disturb nearby residents, but it would also threaten the river ecology. Among the ways the construction company allayed concerns of its human neighbors were to issue grants to cover the cost of sound-reducing windows and doors and to limit the hours of the noisiest activities. To keep waterborne sound waves from harming fish, cushions of aerated water called "bubble curtains" were introduced around the piles being driven so that the underwater noise was abated. Such actions fall into the highly subjective category of "quality of life," an intangible measure of a society's sensitivity and success, but one that relates closely to that most tangible of things: our infrastructure, the part of our built environment that makes it possible to live comfortably and enjoyably in a civilized way.

Even infrastructure buried underground, especially at subway stations, can promote an image of the transit system itself or the area where it interfaces with the aboveground. New York's City Hall station, the southern terminus of the metropolis's first subway system, was famous for its architectural elegance. Designed by the Spanish architect Rafael Guastavino in a Romanesque Revival style, it was lit by skylights and brass chandeliers and had arches made of the architect's famous interlocking terra-cotta tiles. Built on a tight curve, the station itself unfortunately is no longer technically suitable for service in a system that uses longer subway cars. The station suffered the indignity of having its skylights tarred over during World War II and has been closed since the end of 1945. There were hopes in the mid-1990s of reopening the station for tours as part of the New York Transit Museum, but rising fears of terrorist attacks killed those plans. Examples of the Guastavino tiling system do exist elsewhere in working New York—most notably in the Oyster Bar on the lower level of Grand Central Terminal—but the grandeur of City Hall station is likely destined to remain a museum piece "in storage."

When infrastructure is built aboveground and remains exposed for all to see, engineers and planners alike must consider how it might be made as much a positive contribution to the appearance of the built environment as creativity, time, money, and politics allow. Of course, some keen observers of the functional see beauty in the spindly-legged transmission towers that support high-voltage lines over broad fields as well as in the gently drooping electric, telephone, and cable wires that sing beside our streets and roads. One such aficionado is Brian Hayes, whose large-format illustrated book *Infrastructure: A Field Guide to the Industrial Landscape* does for the built environment what nature guides do for earth's geology, flora, and fauna. In his preface, Hayes states bluntly that his "chief aim is simply to describe and explain the technological fabric of society, not to judge whether it is good or bad, beautiful or ugly." But the book's thousand or so color photographs belie his claim of neutrality. The thoughtful composition of his photos, juxtaposing as they so often do the natural and the technological, suggest a complementing rather than a clashing of the given and the made world. His photos demonstrate how still lifes and tableaux of naked technological objects can possess an intrinsic geometric beauty as captivating as the amathematical lines in a portrait by a master painter. Infrastructure not only can be but is beautiful to look at and to use.

The function of a bridge may be simply to carry a road over an obstacle like a wide river or deep valley, but designing the bridge to be an object of geometric and aesthetic interest as well makes it more than a bridge. It can be a thing of beauty, a piece of sculpture, an objet d'art, a landmark. Making a bridge more than a bridge provides for its users a piece of infrastructure that satisfies the mind and elevates the soul. When I first saw photos of the Dames Point Bridge, which crosses the St. Johns River in Jacksonville, Florida, I thought it just another cable-stayed structure, and one that lacked grace at that. However, when I drove over it, I saw it in an entirely new light. I saw the bow-tie shape of the towers' cross members to be a small but brilliant touch for a structure that might otherwise have lacked geometric relief and visual interest. As I crossed the bridge in bright sunshine, the sharp contrast of the shadows of the cables on the bow ties and roadway broke up the masses of concrete and asphalt in a way that was dynamic and exhilarating. Out of the simple forms making up the bridge came a complexity of movement that was perfectly attuned to the time and place in which I found myself. Experiences like that enhance quality of life.

HISTORICALLY, AS HUMAN SETTLEMENTS and their organization became more complex, so did their infrastructures. Getting sufficient amounts of water to centers of population in the Roman Empire was as important as getting soldiers efficiently to its outer reaches. Indeed, the rise of an empire was as dependent upon the quality and reliability of its infrastructure as its fall was hastened by the deterioration of its roads and other means of communication. At the peak of their civilization, the Romans demonstrated their engineering prowess by building aqueducts, which might be considered "water bridges," to bring the essential liquid resource from distant springs to urban centers. The Pont du Gard, which carried water across a wide valley in southern France, stands as a two-millennium-old monument representative of the achievement of Roman concept and execution that made possible fountains, baths, and more—at the same time symbols of technological control and social interaction—that in turn made civilized living possible and enjoyable. Needless to say, structures like the Pont du Gard are also aesthetic masterpieces.

The bow-tie design on the towers of the Dames Point Bridge, which carries Interstate 295 across the St. Johns River in Jacksonville, Florida, provides visual points of interest on what might otherwise appear to be a too-rigidly composed structure. Officially named the Napoleon Bonaparte Broward Bridge, after the early twentieth-century governor of the state, the structure dominates the otherwise flat area through which it carries the city's East Beltway.

The challenge of providing the infrastructure needed to supply water
to a community seems often to have brought out the best in engineers,
architects, and city planners. When the rapid growth of Philadelphia in
the early nineteenth century made a new waterworks necessary, one was
established on the eastern bank of the Schuylkill River, near where the
art museum is today. But rather than being outfitted with functional
sheds housing the steam engines needed to pump the water up to a reservoir, the Fairmount Water Works, whose initial construction phase dates
from 1812 to 1815, was designed with stunningly attractive buildings in
the Greek Revival style. The surrounding area contained a park, which
drew locals and tourists alike to marvel at the technology and revel in
the scenery and in the serenity among the steam engines. The Fairmount

*This 1835 engraving of Philadelphia's Fairmount Water Works on the
bank of the Schuylkill River shows how modern infrastructure can be
built to meet classical architectural standards. The water works had an
adjacent park and gardens, and the site remains a recreational and
tourist attraction. The bridge to the right was designed by the engineer
Louis Wernwag, who also worked on the waterworks. The bridge's
340-foot record-setting span earned the structure the nickname
Colossus.*

Water Works site remains a tourist attraction today, as does Chicago's
1869 Water Tower that later gave its name to a nearby high-rise building
containing a luxurious urban mall.

When New York needed to bring fresh water down from the hills to
its north, the Croton Aqueduct was conceived. The original system,
which was designed under the direction of chief engineer John B. Jervis
and constructed in the 1830s and 1840s, included a dam across the
Croton River and forty-one miles of tunnels and other conduits to the
city to the south, including a reservoir at the present location of the
New York Public Library. Faced with the decision about how to carry
the water across the Harlem River separating the island of Manhattan
from the mainland, Jervis rejected the option of employing a pressurized siphon that could be buried out of sight under the river and instead
designed a graceful stone aqueduct supported by multiple arches in the
Roman style. He produced the High Bridge, which was opened in 1848
as a visible symbol of what the citizens of the city were paying for and
waiting for.

In a 2015 retrospective consideration of the structure published in his
regular *New York Times* column about the city, Jim Dwyer wrote that
the High Bridge was "the greatest public work in New York City's
history," making the modern metropolis possible. Manhattan's reaching
into Westchester County for fresh water was prompted by the over-
pumping and consequent contamination of its water wells with sewage
and brackish water, which led to a cholera outbreak in 1832. Those who
could moved out of the city, and its growth might have been permanently
stunted had Jervis not overseen the construction of the aqueduct,
through whose gently sloping mains as much as ninety million gallons
of clean water could flow southward daily through the force of gravity
alone. Waxing lyrical, Dwyer saw the aqueous journey as "pure,
unfiltered water rolling stanza by stanza through a poem of civil
engineering."

With the continued growth of the city and the concomitant need for
more water, an additional pipe had been added to the High Bridge in the
early 1860s; in the early 1870s a new reservoir was added around
Amsterdam Avenue and 173rd Street in Upper Manhattan. A more

HIGH BRIDGE AND HIGH SERVICE WORKS & RESERVOIR.

The High Bridge over the Harlem River between the Bronx and Upper
Manhattan was part of the Croton Aqueduct. In the early 1870s, the
Highbridge Reservoir, shown here in the background, was added to
the system. The reservoir and the water tower beside it provided pressure sufficient to carry water to buildings located in the surrounding
area. The smokestack marks the location of the structure housing the
steam engine that pumped the water to the elevated heights of the
reservoir and water tank.

substantial expansion of the system begun in the late nineteenth century resulted in the construction of a new dam—then the tallest in the world—over whose stepped spillway the falling water was a visual and aural delight, as could be appreciated from the soaring arch bridge just downriver from the dam. But with the opening of the New Croton Aqueduct in 1890, the High Bridge was no longer an essential part of the New York City water supply system, and it ceased to serve as an aqueduct in 1917. Shortly thereafter the Army Corps of Engineers declared the bridge a hazard to navigation, and in the late 1920s its stone piers in the Harlem River were demolished and the resulting gap spanned by a single steel arch. As with so much once integral infrastructure, the High Bridge had changed with the times.

Though it had never carried vehicular traffic, the bridge was a convenient pedestrian link between the neighborhoods in the Bronx and Upper Manhattan that it connected. However, when in the 1950s some objects thrown from the walkway injured tour-boat passengers, the bridge came under scrutiny and was eventually closed to pedestrians and remained so until 2015, when a restored High Bridge provided a park-like space high above the Harlem River.

IN THE MIDST OF the Great Depression, Conde McCullough, the Oregon state bridge engineer, oversaw the design and construction of a series of graceful concrete and steel bridges along the state's scenic Pacific Coast Highway that stand today as delights to see and use, and provide yet another demonstration of the fact that essentially functional structures need not be aesthetically inferior. Likewise, the Brooklyn Bridge is an outstanding example of how engineering can rise to the level of art. The thoughtful use of stone and steel in the structure combined the best of the old and the new to create an enduring masterpiece.

By comparison, the two bridges just upriver from the Brooklyn offer the opposite experience. The Williamsburg Bridge is one of the most aesthetically challenged structures of the early twentieth century. Heralded as the longest-spanning suspension bridge in the world when it opened in 1903, this ill-proportioned, utilitarian assemblage of stark steel has served the city for more than a hundred years, but without any grace. It is an unfortunate emblem of the purely functional public works project. After the disappointment of the Williamsburg, the Manhattan Bridge restored confidence that an all-steel structure could also soar aesthetically. Opened in 1909, it has stood for a century between the Williamsburg and Brooklyn bridges, as if trying to prevent a direct comparison between the two. Still, however successful architecturally the Manhattan Bridge may be, it has not aged well structurally. Because the subway tracks that run across it are located along the sides of the bridge deck, the structure twists one way or the other each time a heavy train crosses over. This decades-long repeated flexing, aggravated by at least one prolonged period of neglected maintenance and unchecked corrosion, hastened the deterioration of the structure, which required seemingly endless rehabilitation. Neither form nor function alone makes for a well-balanced design that can stand the test of time.

Earthquakes are natural enemies of bridges, and engineers have learned a lot about these ground-shaking events since some of the classic components of our infrastructure were built. The Golden Gate Bridge has had to be upgraded to make it capable of tolerating a major earthquake while maintaining its aesthetic integrity. The instrumentation now inconspicuously mounted on the bridge not only enables engineers to learn from its response during moderate earthquakes but also provides a diagnostic tool to assess whether the bridge is safe to carry traffic immediately after a big one hits. Such proactive forensics can be expected to be done increasingly also in anticipation of possible terrorist attacks on iconic structures. The challenge is to make them stand proud and strong in the face of adversity while retaining an architectural integrity.

When the design of architect-engineer Santiago Calatrava for a Lower Manhattan transportation hub was unveiled in the wake of the 2001 attacks on the World Trade Center, it promised to be an infrastructure project that would help revitalize ground zero in substance and in spirit. An "oculus" above the transit hub's great hall was to be fitted with movable sculptural steel wings that would celebrate form and function, architecture and structure, equally. However, budget considerations eviscerated the design: only a glass skylight opens to commemorate September

11. It is also a reminder of how politics and money can crowd out sensibility and art. If we can forget what might have been, the sheer scale of the space still gives it considerable grandeur. But the $4 billion, ten-year project—at twice the original budget and six years behind schedule—has been described as "one of the most expensive and most delayed train stations ever built." According to its executive director, the Port Authority would not today give priority to the hub over other infrastructure needs.

Contemporary megaprojects, which have been defined as those—iconic or not—that cost at least $1 billion in 2015 dollars, have become common in the United States and around the world. Oxford management professor Bent Flyvbjerg has concluded that megaprojects account for some 8 percent of the global domestic product, with China being responsible for a large part of the total. By one estimate, in its rush to develop its infrastructure, that country consumed more cement during the 2011–2013 period than was used throughout all of the twentieth century in the United States. Flyvbjerg believes that most such projects are either poorly executed or should never have been undertaken. He calls it an "iron law of megaprojects" that they end up "over budget, over time, over and over again," with nine out of ten having cost overruns.

Comparing projects undertaken a century or so apart is not necessarily a trivial computational exercise. The relative value of the U.S. dollar, for example, can be calculated using for the years involved the relevant consumer price indexes; gross domestic product measures; or the unskilled wage rate, which may be most relevant for nineteenth-century construction projects involving considerable hand labor. The Erie Canal, which was built between 1817 and 1825, cost $7 million in the dollars of the time. Today, it would cost $170 million as measured by a GDP statistical deflator index, but $1.74 billion using unskilled wage measures. Thus, by the former conversion, it was not a megaproject by today's billion-dollar standard; by the latter calculation, it definitely was. Perhaps the best way to identify historical megaprojects is to expect that we will know them when we see them or consider descriptions of them.

The temptation to a department of transportation, city planner, engineer, or elected official of leaving the legacy of an iconic or signature bridge, building, or other prominent piece of infrastructure often trumps the risk of overreach, and supporting examples abound, including the Sydney Opera House, Boston's Big Dig, and the East Bay Span of the San Francisco–Oakland Bay Bridge, each of which was wildly over budget and chronically delayed in its completion by unforeseen construction complications or public opposition. The Seattle Alaskan Way Viaduct replacement tunnel project threatened to join such a list, but officials assured the public that the delay caused by the damaged tunnel boring machine Bertha being idle for approaching two years would not raise the cost of the project over the initial estimate of $4.25 billion for the two-mile-long tunnel and associated improvements. As of the summer of 2015 that remained to be seen.

Model megaprojects do exist, and it did not always take billions of dollars to produce them. This is attested to by the majesty of the West Bay Span of the San Francisco–Oakland Bay Bridge, the Hoover Dam, and the Empire State Building, to name a few from the 1930s. Grand concepts and great projects that result in glorious monuments that may last for centuries can show many generations hence what we were at our best. The Works Progress Administration and Civilian Conservation Corps, which were created during the Great Depression, are oft-cited models for what can be done during tough times. Much of what is best in America's national parks and scenic highways stands today as a legacy to the focused human energy that was embodied in such programs. Among the improvements made by the CCC were the construction of trails and roads in Yellowstone and on Skyline Drive in Shenandoah National Park. While there has been much talk about infrastructure in recent years, the focus has too often been on using "shovel-ready" projects to stimulate the economy. This may be politically and economically expedient, but it is not necessarily going to result in any models of engineering sense and architectural sensibility. It is true that there is much to repair and replace in our aging infrastructure, but haste makes waste, and function without form does not uplift the soul.

10

Because It Was

GUARDRAILS, MEDIANS, AND JERSEY BARRIERS

NO MATTER HOW BEAUTIFUL a highway or how iconic a bridge, it will not be safe to drive on or over unless it is properly designed with regard to the geometry of its approaches and its curves. Too steep an incline on a bridge approach will slow down trucks to a dangerous degree, leading to excessive switching of lanes by vehicles, including those that cannot handle the grade. Highways that have not been laid out with smooth transitions between straight and curved sections can be equally dangerous, making it difficult for drivers to keep their vehicle in the proper lane. But even the geometrically well-designed road can be hazardous if the drivers of cars and trucks using it encounter no warning or barrier to their drifting off the road or into oncoming traffic. Preventing this from happening is part of good infrastructure design. The standardization of road geometry and road markings helps us achieve as much safety in driving on an unfamiliar road as on one that is familiar to us. But even the supposedly familiar hometown road can present us with surprises.

Local references to roads and highways often do not reflect what their official names or numbers are. Southern Californians are famous for prefixing highway numbers with the definite article, thus referring to Interstate 105, which carries traffic toward and away from the Los Angeles International Airport, as "the 105." In and around New York City, driving directions tend to be given in terms of names rather than numbers. Thus, instead of being directed to Interstate 95 North, drivers coming onto the upper reaches of Manhattan Island via the George Washington Bridge and heading for New England will be told by New Yorkers to take the Cross Bronx Expressway. In Durham, North Carolina, U.S. Highway 501, which skirts the western edge of the older part of town, is frequently referred to simply as "the bypass." It is coincident with U.S. 15 along this stretch, and many locals designate it neither as the bypass nor as Route 15 or Route 501 but as "15-501," an appellation that can be quite confusing to outsiders.

As I drove north on this road of aliases one morning, I watched for the green informational highway signs that direct drivers to state highway NC 147, also known as the Durham Freeway. Over the years, this road was extended through and beyond the city to connect with Interstate 85, passing over 15-501 in the process. Since the exits on the existing bypass had been numbered consecutively, when new exit ramps from 15-501 onto the extended route 147 were added, they were distinguished from each other and the existing nearby Exit 108 by suffixes, which resulted in the crowded confusion of Exits 108A, 108B, 108C, and 108D—all within a few hundred yards of each other. This also made for a trying traffic pattern, with vehicles entering onto the bypass from Morreene Road (Exit 108A) crossing over lanes of traffic exiting from the bypass and trying to get onto the freeway heading either south (Exit 108B) or north (Exit 108C) as well as onto Hillsborough Road (Exit 108D). It is as confusing to drive as it no doubt is to read about.

The exit I was looking for was number 108B, and to reach the ramp I had to exchange places in a very short distance with the cars coming down the ramp from Morreene Road. This naturally commanded my immediate attention, and so when I first got a good look at my destination exit ramp, I was nearly upon it. The ramp is preceded by a yellow caution sign displaying the advised safe speed of 15 miles per hour for negotiating the sharp curve I was about to enter, but on that day the sign was blocked by the traffic with which I was trying to exchange places. As I entered into the curve, I immediately hit my brakes, because I not only felt its tightness but also saw how badly dented, bent, torn, and twisted the guardrail was in the narrow grassy patch separating the on- from the off-ramp of the freeway. The disfigured guardrail was also badly rusted. This was all clear evidence that vehicles had repeatedly and for some time taken the turn much too fast and had lost control.

As I progressed along the ramp, the notches on the tightly curved guardrail became spaced farther apart, indicating clearly that the number of fenders and worse over which it had claimed a victory declined with distance along it. The damage to the rail was a visual record of how the vehicles going fastest went off the road earliest, and those that survived the first few feet of the ramp had had a chance to recover in the very tight turn. In addition to the safe-speed yellow warning sign that is susceptible to being obscured by merging traffic, another one at the entrance to the ramp would likely have saved many a collision with the guardrail; but a vehicle going maybe way too fast took that sign down, and it had yet to be re-erected. In a subsequent drive along the same route, I noticed a lot of damaged and flattened highway signs that would be easier to see and read if they were fixed and straightened. No doubt the same thing has happened on dangerous roads across the country, and repairing signs promptly would not only improve the safety of our highways but also reduce the ugly clutter beside them.

In the case of the very tight curve of exit ramp 108B, I wondered if the ghastly warning that the badly mangled first twenty or so feet of the guardrail gave to drivers motivated the highway department to leave it unrepaired. It was certainly an eyesore, but it may also have been an inanimate hero. Indeed, without the disfigured guardrail between the ramps, to drivers distracted by traffic the way I was, no other visual indication whatsoever would warn that this is a very dangerous curve and that a driver entering it should definitely reduce speed.

A few days after my close encounter with the mangled guardrail on the Exit 108B ramp, my wife was driving us north on 15-501. Since I was riding in the passenger seat, my eyes were free to wander and survey. What caught my attention long before we reached Exits 108A through D was how many sections of guardrail along the highway proper had been repaired. It was easy to identify the replacements because the newer steel sections did not match in color or texture the preexisting ones. Some sections were bright metal; others dully galvanized; and older rail that had been grazed or lightly dented and left in place had patches of rust, with some sections totally rusted. The overall effect was far from what could be called highway beautification. When we reached Exit 108B, I was pleasantly surprised to see the old guardrail replaced with a bright new one for a good ways along the ramp. I hoped that its galvanized steel would remain untouched and unblemished for years, but I knew that to be an unreasonable expectation, given that the tight geometry of the curve would stay the same. Indeed, on a subsequent trip only a few weeks later, the new guardrail bore scars of serious impact.

THE CONCEPT OF A guardrail must surely have evolved from that inherent in the arrangements of closely spaced boulders found along the cliff edge of long-established mountain roads. While self-preservation may keep animals on their path beside a precipice, wagons and motor vehicles have no such instinct, and their drivers are susceptible to the distraction of the spectacular scenery that usually abounds in such a location. Certain stretches of road, especially those with opposing traffic coming around blind curves, must have gained reputations for being particularly dangerous. Placing large rocks along the outside edge of such a road would obviously provide a measure of safety against wheels inadvertently running off the edge.

An ample supply of large rocks or boulders would naturally have resulted from cutting a road along the side of a mountain, and so the raw materials for a safety barrier would have been conveniently located for use. Just as pioneer farmers made boundary walls out of stones they cleared from the land, so road builders formed edging out of the rubble that resulted from their digging and blasting. In time, the idea of marking the edge of any dangerous stretch of road with virtually immovable objects like heavy stone walls would have been incorporated into good road building practice. Eventually, the idea of lining roads of all kinds with boulders, walls, and other devices to keep vehicles from

Going-to-the-Sun Road is the only highway through Glacier National Park in Montana. Completed in 1932, it is one of the first designed with the touring automobile in mind. The mountain road is bordered at this location by a handsome rock wall, no doubt formed from the spoils that resulted from cutting the road into the side of the slope.

straying from the road surface onto excessively steep slopes or otherwise dangerous roadside conditions would become commonplace.

Where rocks and boulders were not available or easily moved into position, logs could serve as curbs of a sort or wooden fences could be constructed not only to delineate the road but also to provide a measure of resistance against a vehicle going astray. Of course, such barriers would only have been as strong as their weakest link. What that was would naturally depend upon the strength of the timber used and the care with which a barrier or fence was constructed. However, even a moderately well-built fence might be effective against relatively light-weight and slow moving vehicles, which horse-drawn wagons and early automobiles generally were. The modern guardrail that is so familiar along interstate and other highways carrying a high traffic volume is a special type of fence, one that has solidly planted posts and specially engineered beams—most often made of steel—designed to stop or redirect the heavier and faster moving vehicles of today.

WHATEVER THEIR FORM AND however constructed, guardrails gener-ally are supposed to be lifesavers, and had one been in place on the southbound side of the bypass when the Exits 108 were newly created and numbered, it would surely have prevented a fatal accident. Shortly after the new set of ramps was opened to traffic, a patrol car was heading southbound on the common exit ramp, when the officer received a radio call that required him to reverse direction as soon as possible. Not yet fully familiar with the new traffic patterns around the multiple exits numbered 108, he made a quick U-turn by crossing the grassy area separating the southbound shunt road and the southbound highway lanes, evidently thinking that he had crossed a median and was turning onto a northbound road. Unfortunately, he was in fact driving at response speed in the wrong direction in a southbound lane. The speeding police vehicle crashed headlong into an automobile heading in the correct direction; the fatal accident claimed the lives of an innocent family of three. Had a guardrail or other barrier been installed—or the officer been familiar with the new traffic pattern—the accident would simply not have happened.

More conventional and familiar highway designs do separate oppos-ing traffic flows with a variety of barriers to vehicles crossing from one side to the other. Wide, grassy medians landscaped with high berms, stands of trees, or deep swales certainly prevent easy crossover, but some medians are just wide lawns, like the stretch around Durham's 15-501 Exits 108A, B, C, and D. These not only allow a determined driver to make a U-turn at speed, however ill-advised, but also allow an out-of-control vehicle to cross over into oncoming traffic. The resulting head-on crashes are among the most deadly on the highways, and of course this is why guardrails and traffic barriers are so important. But they are also expensive. An estimate made in the late 1990s concluded that even if only 10 percent of the more than four million miles of roads and highways in the United States were fitted with guardrails or median barriers, it would represent an investment of almost $65 billion. The number of lives saved could be incalculable.

Such a large investment would not likely have been made without the aid of the federal government. While the individual states are ultimately responsible for interstate and other highways within their boundaries, they expect to be reimbursed for work done in conformity with federal standards and guidelines. In the case of guardrails and barriers, for example, only commercially available products that have been certified through testing are eligible for reimbursement. This naturally all but mandates that a state department of transportation use certified products, which explains why there is so little variation in guardrails and barriers from state to state.

Where a median is narrow, as it often is around metropolitan areas, vehicles are typically prevented from crossing over into oncoming traffic by vertically tapered concrete dividers commonly known as Jersey barriers. In spite of the name, these reinforced-concrete median structures appear to have been used first in California. In the mid-1940s a version of them was installed on U.S. Route 99 south of Bakersfield, where the highway comes down into the central valley from the Tehachapi Mountains. Using the durable dividers in place of guardrails or other, more transparent barriers against car and truck crossover appears to have been motivated at least in part by a desire to reduce the need for highway crews to make repairs in that dangerous location. The concept of the barrier itself may have been suggested by the high upward curved "safety curbing" that kept drivers in their lanes on the express motorways depicted in the 1939 General Motors Futurama exhibit.

Whatever their origin, rubber- and paint-scarred concrete traffic barriers along highways today demonstrate that they can take a lot of

impact without suffering structural damage and so need little maintenance. Still, like all technology, whether infrastructural or not, early
barrier designs exhibited limitations and therefore evolved over time. In
1955, the first ones used in New Jersey highway medians had vertical
sides rising to 18 inches from a curbed base. These were not fully effective, and—guided by the nature of the accidents that occurred with
them—engineers gradually changed the barrier shape and height, which
in 1959 reached the now-familiar profile with a height of 32 inches.
Variations on the Jersey barrier that have been developed elsewhere
include the Ontario tall-wall barrier, which has a height of 42 inches,
and the F-shape barrier, so called because it was the sixth variation of
profile in a sequence labeled A through F that computer simulations,
confirmed by crash tests in the 1970s, showed to be superior to the
Jersey shape. However, the performance of the F shape did not prove to
be superior enough to justify state highway departments investing in
new concrete forms to produce the shape. The departments felt that
their limited infrastructure budgets could be better spent.

The inclined sides that characterize the shape of the typical concrete
barrier give vehicle tires and bumpers something to ride up on, thereby
converting the kinetic energy of the errant moving vehicle into potential
energy as it rises up the barrier face. This is a more benign way to lessen
the energy of a horizontally moving vehicle than is crushing its parts.
General Motors had developed a so-called GM-shape barrier design, in
which the change of slope of the barrier side occurred farther up the
wall. This tended to lift the wheels of smaller cars that were popular
in the 1970s, such as the Chevrolet Vega, to too great a height, causing
them to roll over. The shape was discontinued after crash tests confirmed
that this undesirable behavior was indeed imparted to impacting
vehicles.

Installing a concrete Jersey median barrier can cost from two to
three times what a steel guardrail does, and so this explains in part
why the latter is more common. However, the opaque concrete barrier,
if sufficiently high, also provides a shield against oncoming headlights
on curves, which is an argument in its favor. Where headlight glare
coming over a barrier top has been a serious problem, devices looking

like vertical venetian blind slats have been installed atop the concrete; but when these become bent or broken off, the overall appearance can be very unattractive and distracting and so reflect poorly on the neighborhood through which the road winds. I recently drove along a stretch of I-95 in Richmond, Virginia, that had long been fitted with these light shields. The line of them had become pretty raggedy, but a newly installed fresh set restored the neatness and effectiveness of the arrangement.

On some especially notable scenic routes, wooden rails or low stone walls can serve the purpose of redirecting drifting traffic. The attractive Merritt Parkway in southern Connecticut uses heavy timbers as what are termed not guardrails but "guiderails," suggestive of the function of the structures. An excellent example of a stone "guardwall" is the one used on the Baltimore–Washington Parkway. In this case, the wall is not solid stone but consists of a stone veneer placed over a solid concrete core; the effect is somewhat artificial looking but quite distinctive.

Another traffic barrier option relies on steel cables, which provide the most transparent and inconspicuous and least expensive alternative to the conventional beam guardrail. Because the cables stretch like rubber bands when a vehicle runs into them, this option can only be installed in wide medians or where there are no solid objects, like trees, too close behind them. Cable systems do have an advantage in snowy regions, for their relative transparency does not allow them to serve as obstacles against which snow can drift or be piled up by plows. They are also easier to repair. Whether spanned by cables or rails, the steel or timber posts employed can be further classified as weak or strong, depending on whether they do or do not yield much to impact. Where they can be used, weak-post systems are preferable because they absorb more energy and so are considered safer.

Heavy-plastic Jersey-style barriers are often used in construction zones. The hollow plastic barrier sections can be filled with water to provide substance and also energy absorption, but the plastic versions do not match their concrete counterparts in overall impact resistance. The plastic barriers do have the advantage of being colorful (typically yellow or orange) and therefore highly visible and conveying a sense of caution near construction sites, around which they can be moved with relative ease.

VOLVO AUTOMOBILES HAVE LONG stood out for their ability to protect their occupants during crashes, and the "crumple zones" integrated into their frame and body designs were advertised as safety features, with the various parts of the car collapsing in a predetermined sequence to absorb the energy of a crash rather than transmit it undiminished to the occupants of the vehicle. Other automobile makers have also prominently advertised that their cars are eminently crashworthy, capable of withstanding an accident of some severity while protecting their occupants. Traditionally, the crashworthiness of an automobile was determined by placing crash dummies in the seats and literally crashing the test vehicle into a wall, guardrail, or other vehicle, but increasingly computer models are being used to establish how well a car and its occupants will fare in a collision. The computer results can be checked by physical crash tests, but not so many of them may be required to confirm the efficacy of a design.

The crashworthiness of a guardrail or other barrier structure provides a different perspective on an automobile impact. The badly mangled guardrail that signaled me to slow down as I entered the entrance ramp to the Durham Freeway is a stark reminder that even stationary steel structures can get badly bent out of shape. And if not properly designed, they can do a lot of harm in an accident. Being too stiff and unyielding is only one undesirable feature. The end of a guardrail, for example, should be designed so that it does not impale a vehicle and its occupants that hit it at high speed. This can be prevented by having the termination of the guardrail slope down into the ground, literally burying its potentially dangerous end. A vehicle striking such an inclined rail can be expected to ride up upon it and maybe turn over, but it will not have its occupants harmed by the structure itself intruding into the vehicle's interior. Other design schemes for keeping the terminal part of a guardrail relatively benign include fitting it with a blunt bent sheet of steel that, when hit, at least at lower speeds, acts more like a spring than a spear. There are also flat-surfaced end fixtures that are designed to push aside the guardrail proper from its posts to keep it from penetrating a vehicle's interior. Sometimes, rather than trying to design benign features into the guardrail system itself, its end is simply shielded by plastic barrels full of sand or water, which can absorb relatively harmlessly the kinetic energy of an impacting vehicle.

The best scenario involving traffic flowing beside guardrails and barriers is one in which the moving vehicles never come in contact with the stationary objects. Unfortunately, even when they are maintaining a safe speed drivers get distracted and daydream; their eyes glaze over and they nod off at the wheel. Sometimes when they veer out of their lane, they are awakened by the horn blast of the vehicle beside them or the rumble of their tires on warning strips just in time to avoid a collision. But the ubiquity of dented and mangled guardrails and skid-marked median barriers attest that not all drivers are so lucky.

11

Grassy and Wanted Wear

STREETS, LAWNS, SPEED BUMPS, POTHOLES

Around austin, texas, where I lived in the early 1970s, Interstate 35 was referred to as the interregional highway or, more often than not, just "the interregional." This terminology, along with the system of secondary roads carrying designations Farm to Market Road and Ranch to Market Road—many bearing four-digit route numbers—was among the linguistic and cultural oddities that made Texas exotic back then to a transplanted Yankee like me. Of course, the term "interregional highway" had been widely used in the 1930s to denote a concept that evolved into what is now known as an interstate. Still, words do have a way of insinuating themselves into a culture like that of Texas, which might be thought of as both a state and a region unto itself.

The naming of roads generally is far from a linguistic science. One etymological theory about the term "highway" is that it originated in the concept of a literally high road that followed a ridge and so benefited from the action of the wind clearing it of leaves in autumn and snow in winter. In river towns and villages, the elevation of a High Street meant that it did not flood in the spring. A more lowly explanation is that the word designated a road that was higher, however slightly, than the area through which it ran and so rainwater flowed off it onto the ground, leaving a relatively dry road surface. The word "highway" is defined in dictionaries today as a main road connecting principal towns and cities, what in the past might have been called a king's highway or royal road (such as California's El Camino Real connecting Spanish missions) or a post road, over which mail and other important information was carried by horseback, on foot, or by word of mouth. In colonial times, a city like New York was connected to other important cities via post roads, the Boston Post Road being the most widely cited. Until the twentieth century and the advent of the automobile, post roads were among the few kinds that the federal government funded.

Synonyms for the term "highway" today include beltway, expressway, freeway, interstate, motorway, parkway, skyway, throughway (often spelled thruway), tollway, and turnpike (or just pike), but the fine distinctions among these thoroughfares can seem elusive and the applications of the words arbitrary. Generally speaking, highways bearing the first handful or so of these designations are toll-free, and the last few charge tolls. Beyond that, the terminology can be somewhat regional, with "freeway" seeming to be favored in California but now used sparingly on the East Coast, where the term "interstate" is more common. In fact, the concept of a "freeway" was espoused in 1930 by Edward M. Bassett, a New York lawyer and president of the National Conference on City Planning, who admired parkways, which he thought of as elongated green spaces to which access was limited. He argued for similar limited-access roads for not only private automobiles but also commercial traffic, including trucks. He chose "freeway" as a "short and good Anglo-Saxon" word connoting not the absence of a toll but rather "freedom from grade intersections and from private entrance ways, stores and factories." As Bassett presented his idea of a freeway in a brief note in the *American City*, "It will have no sidewalks and will be free from pedestrians. In general, it will allow a free flow of vehicular traffic. It can be adapted to the intensive parts of great cities for the uninterrupted passage of vast numbers of vehicles."

Off the highway, what local streets or back roads are called can seem even more arbitrary and odd. The larger and more important of these typically have surnames that sound like parade routes or shopping venues: Avenue, Boulevard, Byway, Concourse, Road, Street. The surnames of lesser roads can sound downright backyardsy: Alley, Chase, Circle, Close, Court, Crescent, Drive, Green, Grove, Lane, Mews, Path, Place, Row, Run, Terrace, Trace, Trail, Way, Wood. Whatever dictionary-fine distinctions exist between the various designations, they seem regularly to be disregarded. The street that runs in front of our North Carolina home is named a Road, but it is barely five hundred yards long and has only about a dozen houses sharing its name in their address. Our Road, as the neighbors call it, is narrow and untouched by centerline or lane markings of any kind. At each end it terminates in a T, running into more genuine and extended Roads, each of which stretches for at least a mile and has a double yellow line down its center. One dictionary I have consulted defines "road" as "a wide way leading from one place to another." Our Road does go from one place to another but it is definitely not wide. Why it was designated Road is a mystery to me.

The given name of our Road, Plymouth, is less mysterious, since all of the older streets in the area carry the names of English towns and counties (Dover, Eton, Windsor, Bristol, Rugby, etc.). A specific name like Plymouth preceding the generic Road or Street is referred to as an odonym, the prefix *odo-* deriving from the Greek word for road or way. Odonyms can also be oddities. The main street in Brunswick, Maine, is spelled not "Main" but "Maine" Street. In nearby Bath, the main street is named Front, perhaps because it could be viewed as the flood-free waterfront, away from which streets slope down toward the Kennebec River. The original city hall building still stands at the corner of Front and Centre streets.

The most common street name in the United States is believed to be not First but Second Street, the "first" street often being named Main or Front Street. I have always appreciated groups of streets named according to a theme, such as states or state capitals. Many towns named their streets after trees, presumably the kinds once planted along them, which may have made street signs unnecessary. Atlanta is infamous for having about ninety of its streets named Peachtree or some variation of the word. When theme-named streets are grouped together, it can help in navigating around town. Employing alphabetical order in naming streets, as sections of Minneapolis do, has the advantage of making it easy to locate a given street in an area. The naming of streets in a new subdivision is usually the prerogative of the developer, and many a one has succumbed to the vanity of naming streets after members of the family, often in no apparent order.

Perhaps I am attuned to the naming of streets and roads because I spent my teenage years in a neighborhood of the New York City borough of Queens where the streets were laid out on a grid and their designations followed an easily discernible rational scheme. The wide east–west streets that carried the majority of the traffic were designated, in order of decreasing volume of traffic, Boulevard, Avenue, and Road. If there were additional but narrower and shorter parallel streets bearing the same number, they were called Drive and Terrace. All north–south streets were designated Street, except the major one that bordered our neighborhood on the west: it was a Boulevard. To the east, the neighborhood was bounded by a Parkway, and only an Avenue or Boulevard crossed under or over it. The layout of streets dictated the numbering system used for houses. Our address was 115-68 229th Street, which meant that it was south of 115th Avenue. The 68 indicated to the initiated that it was about a block south of that avenue, close to 115th Road. In fact, it was the second house south of the intersection, on the west (even-numbered) side of the street. How much more rational can you get?

Although it was located within the borders of New York City, our neighborhood was suburban in character. Grass was everywhere: we had a front lawn, a back lawn, and even a driveway median that had to be mowed. Most of our neighbors' cars were left in the driveway or garage during the week, the breadwinner taking the bus to where he could catch the subway, usually into Brooklyn or Manhattan, where the jobs were. Because so many people commuted to work or school, each morning they walked in droves toward Linden Boulevard to catch the bus, and each evening they walked away from Linden to return home. Each side of each street had a sidewalk, which was separated from the street by more grass. The infrastructure was thoughtfully laid out and appreciatively cared for and used.

On the street in North Carolina where I live now, there are no side-walks or curbs, and the edge of a grass lawn provides a fuzzy line of demarcation between road and not road; but pedestrians and amblers cannot easily walk on the grass, because the lawn drops off—rather steeply in some places—right beside the road into grassy or rock-lined gullies. These diminutive vales continue through culverts beneath driveways and streets. After a heavy thunderstorm, they channel large quantities of rainwater downhill into normally dry creek beds and into storm sewer openings. These latter are few and far between, which means most streets are not interrupted by sewer grates that can get clogged with leaves, complicate paving, and present hazards for cyclists.

There was a storm sewer grate right in front of the house in which we lived previously, and on occasion it presented a major inconvenience. The house was located on a street confusingly (to me) designated an Avenue, even though it ran for only two blocks to connect a major road, which ironically bore the designation Street, and a state highway, which in this vicinity was designated a Boulevard. In any case, our short avenue was located in a neighborhood that had an extensive buried sewer system. Unfortunately, its function was regularly challenged by even a moderate thunderstorm. Our house was on a lot situated at a relatively low section of the street, and that is precisely why there was a sewer grate at the concrete curb in front of the house. By nature and by design, all storm water ran downhill from all directions—from the crest of the road to the side, and from the east and the west to the low between—toward the curb and along it to the sewer grate. The streams picked up and carried with them all the leaves, grass clippings, dirt, silt, and other debris in their way and deposited the tossed salad of shoots and greens atop the grate. It was like city traffic rushing toward a tunnel entrance only to become gridlocked at the clogged mouth. In the street before my house, this action proved to lay a virtually impenetrable covering over the openings in the grate, and so the water spread out into a widening and deepening pool that could rise over the curb and inundate much of our front lawn. The only way to deal with this was to go out in the rain with a rake and remove the soggy mat from the grate and dispose of it a sufficient distance away so that the rushing water draining from the pool did not return the mass of greens and browns to the grate.

These two streets—the ones of my present and previous homes—are located in the same city, but they are contained in two distinct land developments. The first—the one with the drainage ditches—was laid out in the 1920s as part of a country club community. With the wide-open spaces and rolling terrain associated with a golf course, it was natural to design the surrounding streets and their drainage system to carry storm water in open channels leading to just a few buried sewer lines, with overflow from any especially heavy downpour cascading onto the rough and fairways, to be absorbed into the ground after the storm had passed. The second street is in a section dating from a couple of decades later. It is located among more steeply rolling hills that are more densely built upon, leaving no open area in which to deposit storm water. Thus, it has a more dense system of buried storm water sewers guarded by too-effective sewer grates.

As the report cards issued by the American Society of Civil Engineers remind us, infrastructure is more than roads and bridges. What goes over and under the roads and bridges is as much a part of it as what we walk and ride upon. Design decisions made decades or centuries ago can and do have profound impacts on how our roads look and function now and how they might be allowed to function in the future. We and our neighbors consider the grassy area in front of our houses to be ours all the way up to the road, and the deeds to our property specify that it is, but the city retains a right of way or easement to a strip of it abutting the roadway. This means that the city can decide to widen the street, plant trees, construct sidewalks, install pipes, and permit others to erect utility poles, bury fiber-optic cable, and do virtually anything else with that part of our land that is deemed to be in the public interest. In this sense, that strip of grass beside the curb is also definitely part of the common infrastructure.

Generally speaking, trucks use our Road only when a delivery is being made or when a lawn care crew in a heavy-duty pickup tows a trailer loaded down with power mowers and related equipment. Since the road is by itself already narrow, the parked trucks reduce it to a very, very narrow one-lane way. Once, an open-sided truck brought and left a herd of goats hired to munch away on ivy that had taken over an undeveloped lot across the street. At the end of the day the truck returned to pick up the sated grazers. On occasion, there will be a tree surgeon's truck trimming branches away from power lines before a wind or ice storm brings them down on the lines. The trees, which might be considered also part of the infrastructural landscape, can seem butchered after one of these pruning episodes, but most people seem to be willing to accept the aesthetic deficit for what they consider the greater good of uninterrupted power in the heat of summer and the cold of winter. Unfortunately, no matter how carefully pruned, trees and their branches occasionally still do fall across power lines.

Every ten years or so, in the wake of a massive hurricane or severe ice storm that affects a broad swath of the region, the power can be out for as much as a week. There simply are not enough service trucks or crews to handle all the outages, and sometimes our local power company has sent most of its own trucks and crews elsewhere to help other areas that were hit before the storm reached us. This can be frustrating for the powerless, but we welcome the practice when those from elsewhere are redeployed to help us. It all depends upon the track of the storm and the extent of the damage it does. During the days without refrigeration and nights without lights, we have time to think about how power outages could be almost totally eliminated if only the exposed and vulnerable power lines were taken down off the poles and buried beneath the easement. The question seems always to be asked after a bad storm and always to be answered the same way: It would be too expensive to bury all the lines now. It can be done cost effectively in new developments, before gas, water, and sewer lines are buried, trees planted, sidewalks poured, driveways installed, and lawns started. Besides, that cost can be passed on to the pioneers who will be moving into the new houses, perhaps attracted to them precisely because of the buried utility lines and so willing to pay the premium for them.

MANY RESIDENTIAL DEVELOPMENTS HAVE had their streets, avenues, and roads laid out to minimize through traffic on all but the major ones. The road network through our 1920s development was certainly designed with this in mind. The one major road, which extends in length only a little over a mile, was deliberately made narrow and curvy to slow traffic and discourage use by other than those accessing roads within the neighborhood. Still, in recent years cars had been traveling at speeds well in excess of the posted limit, so those residents along the road who complained persistently got speed control devices installed in front of their property.

Not all speed control devices are deliberately installed, of course, and potholes—especially large ones—have long served to slow down traffic of all kinds. The word *pothole* itself predates the invention of the automobile. According to one etymology, *pothole* comes from combining the Middle English *pot*, meaning pit—especially one resulting from mining or peat-digging—and *hole*, making the word a redundancy. The *New Oxford American Dictionary* on my desk gives as the first meaning of *pothole* "a deep natural underground cave formed by the erosion of rock, esp. by the action of water." The word also denotes "a deep circular hole in a riverbed." Other geological meanings relate to hole-like depressions in glaciers and gravel beds.

As the word relates to infrastructure, my *New Oxford* defines *pothole* as "a depression or hollow in a road surface caused by wear or subsidence." This meaning is generally dated, like the automobile itself, from the early twentieth century. Of course, what we now call potholes must have been no less annoying to riders in wagons or on bicycles, who wanted better roads as much as early automobile drivers did. The speed bump and hump are later inventions, and the etymologies of these terms are transparent.

Whatever they are called, these impediments that can make drivers curse an otherwise smooth road naturally cost money to install, and this often must come from the same budget that is used to fix potholes. Thus, there develops a classic infrastructural dilemma of choice: to spend money deliberately raising for a complaining individual or small group an otherwise undesirable bump in the road—or to spend it for the good of all by filling unwanted holes in the pavement. In all likelihood, which gets priority will depend upon the timing and persistence of a complainer as well as on the mood and discretion of the bureaucratic public employee involved. However it happened, the principal road through our development now has a total of seven speed humps along its approximately one-mile length.

From one point of view, speed humps can be ultimately unnecessary. If a road were left entirely without the humps and not kept in good repair, over time it would develop potholes. If the potholes were not filled promptly, they would grow in size and likely multiply and coalesce in the neglected pavement. The effect of these growing potholes would be to slow traffic in much the same way speed humps do. In other words, a community could save a lot on road maintenance if it did not rush to install speed humps or to fill potholes at the first complaint. Of course, a proliferation of potholes would present an eyesore in a neighborhood of neatly trimmed lawns and well-maintained houses, and so this eminently practical solution to traffic control is seldom if ever given a chance, at least in better neighborhoods. The pothole is seen uniformly to be a blight on the roadscape in a way that the speed hump, curiously, is not.

We know exactly where speed bumps and humps come from because they are deliberately designed and installed. But understanding the genesis of potholes, the antimatter that no one wants to encounter, requires a deep view of the nature of pavement and its foundation. According to this model, the problem begins beneath the surface of the blacktop, in the very base on which the road rests. Rainwater and spring melt seep into the ground and find their way beneath the roadway. This does little harm when the temperature of the road remains above freezing, but when it falls, the water beneath the road surface solidifies into ice. Since ice takes up more space than the water from which it forms, the volume that was occupied by the water is no longer sufficient to contain the ice, which then presses against the bottom of the roadbed and the asphalt above it. This may cause just a slight lump in the road over the ice pocket, but since the rising lump stretches the surface of the asphalt the way an inflated balloon stretches the rubber of which it is made, the paving material is pulled apart and small cracks develop in the surface.

Over time, small cracks grow into larger ones for a couple of reasons. Water collects in the cracks and freezes, whereby the ice expands and widens the openings. Wider cracks fill up with more water, and the cycle repeats, ratcheting up the width and depth of the cracks still further. In the meantime, traffic continues to use the road, and the repeated passing of tires over the cracks alternately pushes down on the road and lets it recover from the pressure. (Only rarely do potholes appear in properly constructed concrete roadways because the concrete is reinforced with steel rods or mesh, which holds together any cracks that develop and prevents them from widening.) The cyclic loading of an asphalt pavement is akin to what happens to an airplane's fuselage when it is alternately pressurized and depressurized with each flight, and the consequence of such repeated loading and unloading is what initiates and grows cracks, a process known as fatigue. A roadway, after thousands of such reversals, gets as brittle as the mortar in an old stone wall and begins to crumble apart.

As the cracks in the roadway widen and coalesce, holes develop. This can happen because the surface material simply breaks up and provides no cohesion, or because the ice that had heaved the pavement up has melted and drained away, leaving a void into which the weakened asphalt falls. Once potholes begin, they will grow by a more direct mechanism. Automobile and truck tires will roll off one edge of a hole and impact the other edge of it, thereby chipping off some of the asphalt binder and aggregate. After a while, the hole will enlarge to such a size that the tire rolling across it will impact the bottom, dislodging even more aggregate and deepening the hole still further. This is why a good-size pothole will have a substantial amount of asphalt debris sitting at the bottom of it.

The more traffic there is traveling over potholes, the faster they will grow and multiply. The greater the speed with which the traffic passes, the faster the process will proceed. The heavier the vehicles hitting the potholes, the faster they will grow to dangerous proportions. This insidious progression is evident every spring to commuters following a fixed daily route to and from work. They can see the potholes expand in size from day to day and eventually run out of unpocked roadway onto which to swerve to miss the biggest of the holes. It is at this point that automobile drivers seek alternate routes. Depending on posted road restrictions, truck drivers may not have that alternative available to them, and so they are left to continue to drive over the potholes and worsen them even further.

Because of the absence of potholes and speed bumps, there is more traffic on our short street than should logically be expected for one that runs such a small distance. However, not only does the principal road, which roughly parallels ours to the east, have speed humps, but so does the street that exactly parallels ours to the west, and both can develop potholes. In other words, we are a bumpless middle way that appeals to drivers who care for their cars, their comfort, and their time. By taking a detour onto our street, at least two speed humps are avoided, potholes are avoided, and time is saved. Thus, speed humps and potholes can teach us something about infrastructure and its discontents: the condition of infrastructure affects the behavior of people just as surely as people affect the condition of infrastructure.

12
—

The Passing There

SIDEWALKS, CURBS, GUTTERS, HORSES, PAVEMENTS

THE SIDEWALK IS A pedestrian object and subject. According to Joan
DeJean in her book *How Paris Became Paris: The Invention of the
Modern City*, the sidewalk is "the most visible sign of the pedestrian's
place in the modern city." She dates the first sidewalks as we know them
to those on the early seventeenth-century Pont Neuf. Although perhaps
unremarkable today, when the "New Bridge" was indeed new, it was the
highly elevated pedestrian spaces along its sides that, in addition to the
absence of occupied structures lining it, distinguished it from the old
bridges of Paris and London. A century after the bridge opened, its
space reserved for pedestrians was considered a "new convenience" that
was "unfamiliar to foreigners." The sidewalks on either side of the Pont
Neuf were accessible via the four or five steps that separated them from
the roadway, which put the pedestrians at a height all but impossible to
be reached by an errant wagon. In time, the levels of sidewalks and road
surfaces in cities everywhere converged upon each other, but to this day
they are distinguished by a difference in elevation, though now typically
only one small step up from the street.

Paris did not extend the sidewalk concept beyond the new bridge
until 1781, two decades after the Westminster Paving Act of 1762, which
redefined the nature of the streets in that section of London. In particular, the drainage channel that had typically run down the center of a
street was to be replaced by a gutter backed up by a "kerb"—as the
British spell "curb"—on either side of it. The kerbs, of course, edged

sidewalks. In many modern manifestations, the curb flows smoothly into the gutter, whose edge can be as weakly defined on the roadside as the waterline on a sandy beach. Where it borders the sidewalk, however, the curb—especially if made of granite—can be as sharply delineated as the Hudson River is from New Jersey proper by the stark and tall basalt cliffs known as the Palisades.

According to traffic control pioneer William Phelps Eno and Charles J. Tilden, president of the Eno Foundation for Highway Traffic Regulation, writing in 1935 in Bulletin No. 1 published by the foundation, highways were being built so rapidly in America that their development had "proceeded with little regard for one most important factor," which was "the provision for pedestrians." Eno and Tilden were referring mostly to suburban and country roads, noting that "the entire width of some highways is taken up by the roadway and on others what is not needed for roadway is left ungraded or so rough that it is useless for pedestrians, equestrians or cyclists," a condition that they felt should not be permitted. They advocated that "a sidewalk or reasonably well made footpath should be provided on one side at least of every highway." Because of the lack of sidewalks or safe roadside paths, according to Eno and Tilden, people were walking less along country roads.

Of course, this was not the case in urban settings, where sidewalks were ubiquitous, as were curbs. When young children played in the runoff following a thunderstorm, the curb and gutter were all that mattered in the street. The curb bent the rush of water flowing down from the crest of the street pavement into a torrent heading toward the grate of the storm sewer. As a child, I played in such an urban environment with as much adventure and enthusiasm as I imagine my rural counterparts might have had in the rocky rapids of a mountain stream. With my friends, I floated rafts woven out of Popsicle sticks downstream, trying to crash through the dam we, as busy little beavers, had built out of street debris. Unschooled in hydrodynamics or the savvy of the wild, we watched the raft slow and spin in the pool created by the dam, the water itself protecting it from a doomed assault by a city boy's natural naïveté. The flow of the water toward the street-side end of the jetty carried the weave of sticks safely outward bound and around.

The gutter was also a sluice in which we panned for gold with our bare hands, seeking prizes that may have been dropped upstream from holey pants pockets or Cracker Jack boxes. One unseasonably warm winter afternoon, when the melting snow had turned the gutter brook into a freshet, my brother and I were fishing in the rushing current. Being two years my junior, he was still cautious about getting too close to the edge, but I was squatting on the brink when I caught out of the corner of my eye something as pale green as the early spring. It floated on the rippling water like a magic carpet on the breeze, and as it approached my fingers dangling in the stream like a teenager's legs from a dock, I saw that it was indeed something magical: a dollar bill! I fished the find from the flow and flattened it out on the top of the curb to dry. It was as bleached as a much-washed pair of dungarees, but it was as good as a fortune to me in the age of penny candy.

In the summer, when the gutter was dry and we children were older, we played marbles in it with the curb as a solid backdrop. The gutter was usually pretty clean, especially after the street sweeper—not the noisy tank-like vacuum cleaners with large rotating brushes that crawl along the curb today, but a man employed by the city literally to sweep the streets—had passed through. Quietly he pushed a cart carrying a pair of large trash cans. The empty one held his equipment: a broad push broom, a straw broom, a long-handled shovel, an upright dustpan, and a broomstick handle terminating in a sharp spike with which he could spear discarded candy boxes, cigarette packs, and other bulk litter—all business end up. He would angle his cart into the curb to keep it in place and, beginning from a good twenty or thirty feet on either side of the cart, use the push broom to sweep the accumulation of dirt, debris, detritus, and dried dog doo into a pile beside the cart. Passing cars and trucks would have already blown the center of the street clean, and so his job was pretty much confined to the gutter, where everything collected. After using his straw broom and shovel to transfer the pile of litter into the trash can, the sweeper would roll his cart farther down the street and repeat the sweeping chore. The street sweeper was often followed by a truck that sprayed water with such force across the width of the street that it washed it and its gutters clean.

CURBS AND GUTTERS HAD long to deal with more than just water, litter, and dog doo. In 1900, city streets and horses went together, but not well. For example, in New York City at the time there were 130,000 horses, each of which produced an average of about twenty-five pounds of manure and a quart of urine daily, much of which was deposited on the streets and found its way into the gutters. During the winter, licensed gatherers and haulers of manure dumped it onto piles beside the river, which was frozen and so could not receive it. Sometimes these piles stayed in place into the spring, attracting flies and breeding disease. Estimates suggest that in New York alone as many as twenty thousand people died each year from the resulting health hazards.

The streets posed other problems as well. Cobblestones set on end in sand or gravel and bound together with mortar had provided early street pavements that were neither overly muddy in wet weather nor very dusty in dry, but the smooth round surface of each individual river- or stream-polished cobble did not make for the most comfortable of rides. Additionally, cobblestone streets, especially when wet, were slippery underfoot. This was a hazard for pedestrians and horses alike, and the sound of horses' hoofs, shod or not, and iron-rimmed wagon wheels passing from one stone to the next made for quite a racket. Veterinarians recommended, but few teamsters followed the advice, that horses working on cobbled urban streets be fitted with rubber-covered shoes, not only to provide for the comfort and safety of the animals but also to make for a quieter and more peaceful street.

To improve on the prevailing conditions, granite cut into prismatic blocks began increasingly to be used as stone pavers, especially after advances in quarrying lowered their price. Granite streets were also less expensive to install and maintain. Some urban horses were even fitted with iron horseshoes that featured a pointed extension or cleat (known as a calk) capable of gaining some purchase in the narrow grooves between the paving blocks. This gave the animal the ability to push back on the blocks without slipping and thereby be able to pull heavier wagon loads. However, horses always begat horse manure. Not only did the dung possess a foul odor and attract flies, but the continual action of iron horseshoes and iron-rimmed wheels on a hard granite surface was like the operation of a pestle in a mortar with regard to turning dry horse manure into a fine powder known euphemistically as "organic dust." Once pulverized, it was blown by the wind to the side of the road and through open doors and windows, thereby presenting an immediate health hazard, to say the least.

The installation of sidewalks, curbs, and gutters kept the road at bay, but only marginally addressed the problem of pulverized horse manure. Using asphalt to pave over cobblestones or pavers made for a smoother surface that, without depressions or crevices, was also easier to clean. However, the only sure way to eliminate the problems with manure seemed to be to eliminate the horse itself. Fortuitously, the automobile drove onto the scene, but it took decades for the transition from horse-drawn to motor-powered vehicle to be completed. In 1900 there were a total of only about eight thousand automobiles in the entire country, whereas the horse population was 13 million. Over the next decade, the number of automobiles rose to almost half a million, but the horse population nearly doubled to 23 million. Trucks lagged behind in numbers, with none registered until seven hundred were in 1904, when there were fifty-five thousand automobiles, but they were both destined to overtake the horse.

As they remain today, from the beginning trucks were much harder on roads than either horses or automobiles. The heavier loads carried by early trucks meant that they could not be fitted with the limited-capacity pneumatic tires of the day and so were equipped with solid rubber ones. These not only made for a rough ride but were also very hard on dirt, macadam, and asphalt-paved roads alike. Around 1912, of the country's more than two million miles of roads, only a bit more than 8 percent were hard-surfaced, and the bulk of these were located in urban areas. The outbreak of World War I created a sharp increase in demand for trucks, and roads across the country suffered accordingly.

Eventually, tire manufacturers did develop pneumatic tires suitable for use on trucks, but by then much of the damage to pavements had been done. No wonder a young Dwight Eisenhower and his Army convoy experienced such poor roads in the wake of the Great War.

Old photographs of highly congested city streets in the early twentieth century show horse-drawn wagons, automobiles, trucks, and trolleys coexisting. Road designers naturally had to take all such modes of transportation into account. Though as late as the 1940s, horse-drawn wagons were still being used commercially on Brooklyn streets by fruit and vegetable vendors and rag purchasers, the displacement of the horse by the automobile was considered a technological godsend; only when the evils of a different kind of air pollution became evident may some environmentalists have yearned for simpler times. But until all cars and trucks are green, cities will likely have to accept the existence of undesirable emissions the way they do the need for providing parking accommodations in lots and garages, under parks, and along curbs.

TODAY, THE CURB IS a highly evolved piece of infrastructure, which is not to say that it does not admit variation in form and substance. In addition to the sheer-sided palisade form, there are curvilinear curbs that provide a more gradual transition in elevation from street to sidewalk. These are more often found in suburban settings and at the edge of a parkway road surface separating it from a grassy shoulder. As with everything else, however, there are pluses and minuses to each species. The curb with a sudden vertical demarcation has the advantage that liquid cannot easily slosh over it onto the sidewalk, nor can slow-moving vehicles easily mount it, but newly quarried granite curb stones can be a hazard to tires, cutting them and causing blowouts even in the process of just parking curbside. After the village of Chatham, New York, installed some new granite curbing in 2009, as many as one hundred automobiles had their tires damaged when they grazed the curb or drove over it. In time, as the curb edge is worn and rounded by snowplow blades and other abusive treatment, the incidence of tire damage will

This 1909 photograph, which appeared in Leslie's Weekly, *shows that the automobile had by then largely pushed aside and bypassed the horse, at least on New York's Fifth Avenue. The policeman halting the carriage in the foreground symbolizes the end of the old means of transportation (and pollution) and the beginning of a new era. Note that there is not a traffic light or other nonhuman means of traffic control in sight.*

decline and the granite curbs will contribute to the solidly established look of Chatham's downtown.

Earlier in the twentieth century, especially where granite was not readily available, concrete was preferred for curbs, including in the convenient integrally precast form of combination curb-and-gutter units, which often included a steel or cast-iron "guard-strip" or "rub-strip" to shield the concrete curb from wheel impact. Today's concrete curbs, which can also be extruded like cookie dough, have the advantage generally of being less expensive to install than granite, but unshielded concrete curbing is subject to more damage than just having its corners rounded. Like poorly placed concrete generally, it can be subject to cracking and spalling, exposing to water and hence to corrosion whatever steel reinforcing it might contain, presenting a weakened and unsightly condition that calls out for repair. The relative costs of concrete and granite curbing are complicated by their differing lifetimes. While some properly cast and maintained concrete may last as long as some granite, a natural stone curb can remain in place on the order of a century, thus outlasting most concrete subject to the same wear and tear. Furthermore, slabs of granite used for curbing can be reused for that or another purpose. Like most infrastructure decisions, the answer to the one of concrete versus granite curbs depends upon

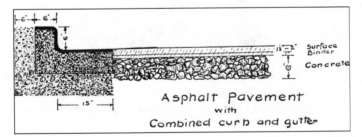

Integrally cast concrete curb-and-gutter units have been used at least since the early twentieth century. The surface of the concrete was sometimes protected against damage by a cast-iron or steel shield, as indicated by the heavy line in this cross-sectional drawing from the fourth edition of City Roads and Pavements, *published in 1909.*

how farsighted a village, town, city, or any municipality wishes to be—and how much money it wishes to or can spend at the moment.

On my university's campus, I walk by a bus stop that is very heavily used when classes are in session. The concrete curb at the bus stop and around the adjacent traffic circle offers a gentle transition between asphalt road and concrete sidewalk. This was no doubt done deliberately so that the many light delivery trucks that park around the circle daily can easily—in fact, are encouraged to—pull partway off the street and onto the sidewalk so that they do not block the cars, trucks, and buses that drive around the circle. (No matter that vehicles so parked encroach on pedestrian space.) The welcoming curb accomplishes this end nicely, but at the bus stop it allows the heaviest vehicles using the road to run their wheels over the curb, the better to park parallel to the sidewalk before letting off their passengers. Where this overriding of the curb has happened repeatedly in approximately the same place over the course of many years, the curb has been crushed open, uncovering the aggregate and steel reinforcing inside. It will not be long before this blemish grows into an ugly rusted and irregularly shaped curb cut that will not be easy to repair in a seamless way. Had the designers of this curb thought about the heavy buses riding over it, they might have made it more substantial, perhaps even specifying a granite or ironclad curbstone.

A curb or gutter is seldom needed with rural roads, especially those raised slightly above the prevailing farm- or marshland. It is better to let rainwater run off the road surface and onto the surrounding fields and marshes. Infrastructure should always be matched to its context. It was only when I became a devoted student of the quotidian that I began to wonder where such delineating but flawed things as gutters and curbs came from and why they were formed the way they were. The gutter, it seems, may have created itself, for all roads have edges or sides that define them, albeit sometimes vaguely. As early roads followed the ups and downs of the terrain, spring- and rainwater naturally followed the pull of gravity and flowed downhill. Over time, especially in steeper terrain, the flowing water would have eroded ruts beside the road, producing favored channels for future freshets to follow. The leap from

such natural topography to deliberately designed and constructed channels and gutters in urban contexts could not have been great.

THERE IS ANOTHER CONTEXT in which curbs and gutters are usually
not needed or wanted, and that is at the private driveway, which often
doubles as a walkway to the house. This is true of our home in North
Carolina. The house faces east on a corner lot that slopes gently downward from the south and west, which of course means that when it rains
at least some of the water flows along the driveway that leads to the
street to the north. After heavy or prolonged rainstorms, the groundwater continues to seep out from the elevated land behind the house for
days, runs onto the drive, and flows down to the street. Like most driveways, our blacktopped one was not laid out according to any department of transportation or AASHTO standard, which means that the
slight curve it takes about two-thirds of the way up to the house follows
no uniform radius or transition. While driving from the street into the
carport area is rather straightforward, even for the first-time delivery-
or repairman, backing out of the driveway presents a counterintuitive
steering challenge. Roughly every fifth car or truck that drives in without
incident ends up backing off the blacktop and onto the grass, and when
it is soggy leaves deep and ugly ruts.

The surface of the drive is covered with asphalt, and despite touchups
it has grown pale gray with age, approaching the color of my hair. The
pavement must be well over fifteen years old, and in one area has raised
welts where tree roots have tunneled underneath it, probably seeking
the water coming down the hill and making this part of the driveway
look like a lawn infested with moles. Where the roots of the tree have
grown especially thick, behaving like frost heaves, the bumps have
expanded the asphalt to the point where it has cracked open. The development of potholes is sure to follow. My wife and I take out one or both
of our cars on average once or twice a day, and a light truck comes up
the drive perhaps once a week. Fortunately, no one drives faster than
about five or ten miles per hour on the driveway. This means that
whatever happens to the pavement, it will progress slowly. However,

someday we will have to deal with this deteriorating infrastructure, and we will face choices and decisions akin to those faced regularly by a town or city's road commission or street department. Shall we resurface the drive with asphalt or convert it to a concrete one? Whichever we decide, how drastically will we attack the existing driveway surface in preparation for laying down the new? Will we have the tree roots removed at the risk of destroying the trees? How thick a pavement of asphalt or concrete should be specified? And what grade or quality of the stuff shall we use? Basically, we will have to decide how much our budget allows us to do.

Because the front of our house faces a narrow road, when we or neighbors have guests, parking can be quite a problem. If cars park with all their wheels on the road, then it is narrowed further to the point where other traffic can have a difficult time getting by without driving on the grass on the other side of the street. As a result, most guests know—or are reminded by their host—to pull off the road surface somewhat (but not so far as to drive into the drainage channel), meaning that where this can be done safely, two tires are on the grass. If the ground is wet and soggy, undesirable ruts will be left in the lawn. To deal with such realities, many of the neighbors have built small, circular drives around which guests can park their cars off the road and on solid pavement.

Our semicircular drive is paved in asphalt, which over the years has developed some common and uncommon problems. Among the first to be noticed was a wide swath of discoloration suggestive of rust. It turned out that the gravel used to seat the flagstones on the walkway leading up from the drive to the house had a high iron content, and so when groundwater flowed down the slope beneath the walkway steps, it carried iron with it, depositing the leached mineral on the circular drive. To fix this problem, we had the affected section power washed. This was done with too much power, however, and so stripped off an inch or so of asphalt, leaving an obvious and unsightly depressed area. The power washer admitted his error and patched the depression to bring its surface up to the surrounding one, but the match was poor and has left a blemish. A new asphalt patch seems to be as difficult to match to the old as is a concrete one, in part because pavements change color with age.

In another area of the drive, there is a shallow, almost rectangular pre-sinkhole-like depression measuring perhaps three by six feet, which I believe marks the location of a buried oil tank that once fed a furnace in the house's crawl space. The location of the depression close to the road suggests a tank positioned for easy filling from an oil truck. Now it is likely gone or rusting, either possibility allowing the soil to settle, probably being accelerated by the groundwater running down the slope. The depression in the asphalt fills with water after a heavy or prolonged rain, and to my eye the puddle is deeper than it used to be. I imagine that a sinkhole could develop someday.

I doubt, however, that my projected sinkhole would ever grow to the size of the one that in early 2014 appeared suddenly in the middle of the display floor in the National Corvette Museum in Bowling Green, Kentucky. That forty-foot-diameter, thirty-foot-deep hole developed when the roof of one of the many caves in the vicinity collapsed. The hole swallowed eight of the twenty-seven prized automobiles on display at the time. The natural disaster did have a silver lining: museum attendance increased almost 60 percent. So instead of plans to fill in completely the accidental tourist attraction, it was to be stabilized and a small bridge built across the hole for visitors to view some reconstructed Corvette carnage below and look beyond it to catch a glimpse of the culprit cave. I doubt that the car of a visitor to my home would be swallowed by any sinkhole that might appear in my driveway, and if it did, our popularity as hosts would surely decrease. Until I can get a definitive assessment of what is causing the deepening depression, it does not make sense to repave the circular drive, but we will likely have to do so before too long. Depending on the details of what is happening underground, we may or may not expect help from the federal Leaking Underground Storage Tank Trust Fund, which was established by Congress in 1986 to help pay for the cost of cleaning up petroleum releases.

When it comes to infrastructure, government responsibility usually ends where the street turns into the driveway. Recently, a funny thing happened on the way to completing the widening of College Avenue in Santa Rosa, California. The elevation of the roadway relative to existing businesses fronting the street was misinterpreted on plans, putting the road pavement as much as three feet below that of the parking areas in front of the stores. Because of the mistake, some driveways between the street and the parking spaces would have needed as much as a 60 percent grade—more than twice as steep as the hillside on which San Francisco's famous eight-switchback Lombard Street is built. The Santa Rosa problem was brought to the attention of Caltrans by the service manager in the College Avenue Midas auto repair shop when he noticed where a new sidewalk was being located. Since storm drains, power poles, and other utilities had already been installed at the lower elevation, the project was put on hold to evaluate what had to be done to correct the error. The cause appeared to have been confusion between U.S. and metric units of measurement: Caltrans had adopted the latter in 1993 for its calculations in response to federal pressure, but reverted back to the use of U.S. units in 2006. The road-widening project had spanned the transition, and so design calculations done in metric units were interpreted at construction time as U.S. ones. If the road were not raised, a lot of cars would have their new muffler damaged as they left Midas unless they drove very, very slowly—as slowly as some infrastructure projects seem to advance. The embarrassing College Avenue error was actually corrected relatively quickly by redesigning the road to be as much as a couple of feet higher where necessary, with construction resuming after only a two-month delay.

EXACTLY WHERE THE LINE between road and driveway occurs may depend upon a range of laws and ordinances and regulations and details relating to local easements and policy. And exactly where along the driveway the transition occurs between private and public use can depend upon values, and ethics, and common sense. In a healthy society full of healthy people, there should be an almost seamless continuity in attitude between what it means to be a good neighbor and a good citizen. In a nation where one comes out of many, there should be enlightened policies about taxes and other obligations, and good citizens should be as willing to pay their fair share as good officials and bureaucrats should be willing to give their fair assessment of property and evenhanded appropriation of services. Wherever the driveway meets the road, a continuity of pavement is certainly desirable.

In the early days of the nation, abutting neighbors, under the watchful eye of a county road supervisor, often worked with rakes, hoes, shovels, and the like to restore a dirt road in poor condition to some semblance of order. By spending "a day on the roads," citizens worked out their taxes for another year. The work was obviously not professionally done, but the practice continued in some states into the early twentieth century. As late as 1913 in Alabama, taxpayers could work off their revenue obligation by spending time laboring on the state highways. With the coming of the bicycle and later the motor vehicle, road supervisors were gradually replaced by road commissioners, who contracted out the work to professionals. While such changes may have made for better roads eventually, the highly political nature of the positions opened them up for abuse in the form of graft and corruption.

13

Had Worn Them Really

QUALITY, SHODDINESS, AND SURVIVOR BIAS

WHETHER WE SET OUT to build a highway or a house, its construction can be done well or poorly; and, when finished, its upkeep can be tended to or neglected. Regardless of the specifics of a structure, its story can exemplify general principles. In the summer of 2014, the op-ed page of the *New York Times* carried an essay of mine in which I treated a piece of private real estate as a microcosm of the nation's infrastructure and lamented the decline in the quality of construction in both. I argued that thinking about the building, aging, and care of small-scale domestic infrastructure, such as a house and driveway, can provide insight into how we as a nation might better respond to our mounting public works problems.

I compared the solidity and quality of original construction both in our home in North Carolina and in our summer house in Maine—each originally built in the early 1950s—to some disappointing improvements that were done on these same structures in the 1980s and 1990s. I lamented the fact that the true two-by-four stud of the earlier era was no longer a stock item in a lumberyard or home improvement store. What is called a two-by-four today is a half inch smaller in each cross-sectional dimension, which complicates repair and renovation in older homes. Furthermore, whereas the workmanship in each of the original structures was superb, that done just thirty and forty years later showed an evident decline in materials used, care taken, and standards expected and accepted. I expressed the same disappointment with changes in commercial construction, and I attributed the decline in part to developers, lenders, builders, real estate agents, and home and institutional owners, who, in a desire to make or save quick money, have created a stock of domestic and commercial infrastructure that ultimately constitutes wasted resources and will not last.

My experience, admittedly anecdotal, led me to suggest that if my observations were shared by millions upon millions of homeowners and apartment dwellers alike, then our conclusions might easily affect what we as a nation expect and accept in public works and how we view our broader infrastructure generally. The short-term fixes and shoddy workmanship we have seen done in our own homes and apartment buildings leads us, by extension, to resign ourselves to the fact that the same inferior attitude and approach is being applied to the construction and maintenance of our roads, bridges, and other forms of public infrastructure.

Granted, we should not have to expect or accept substandard work done on the property we own, but unless we are able to and willing to spend top dollar—if indeed even that can buy quality materials and workmanship—experience and conventional wisdom have conditioned us to think that we have no choice. A homeowner can argue with a contractor only so long before giving up on having something done correctly after two or three attempts. Engaging lawyers and taking the matter to court has to be the option of last resort, and one that hardly guarantees getting the job done to our satisfaction.

To the homeowner who has given up on getting a recalcitrant contractor to right a wrong, it may not be surprising that hastily repaired potholes can reappear in weeks, if not days; that a newly repaved stretch of road can look and feel like a washboard; that a bridge that seems to be perfectly serviceable is being replaced when the road leading to and from it appears to be in worse shape; and that it seems to take forever to complete a highway project. It is no wonder that the voting public, and by extension their representatives in Congress, lack enthusiasm for raising taxes to fund infrastructure projects that they might see as throwing good money after bad.

I concluded my op-ed by noting that we do not have to be citizen-craftsmen who work on our own homes to know that things need not be this way. And we do not have to be homeowners or highway engineers to know that good materials are better than poor and a job done well from the outset will outlast one done badly. As we debate how to pay for infrastructure, we should also have a discussion about raising expectations for what we are buying. Homeowners, project managers, and legislators alike should want to call to account suppliers and contractors who do not produce the quality of materials and work they promise. A roof or road that does not meet agreed-upon standards should not be accepted; it should be redone—at the irresponsible party's expense—until it is done right. Such challenges will naturally lead to delays and legal proceedings, but that is the price for getting things done right. In time, one would hope, doing it right the first time will once again become wise and common business practice, and we can look forward to infrastructure that looks good, works well, and lasts.

The short essay elicited hundreds of comments, ranging from sympathetic agreement—complete with confirming examples—to screeds about my being an academic in an ivory tower, an elitist, and a fogey hopelessly yearning for the good old days. Correctly, some commenters accused me of generalizing from anecdotes, with no valid database that houses were built better a half century ago. Many readers made some excellent points, among them being that technology generally has improved over the decades, as exemplified by automobiles today being safer and more reliable than those of my perceived golden era; that houses are more than just structure, and newer ones generally have more robust electrical systems, draft-free windows, and energy-efficient appliances; that nature did not produce enough quality materials to supply every house for every family in America, let alone around the globe; and that the decline in careful and caring workmanship coincided with the decline in influence of union apprentice and journeyman systems. Many of the comments also had to do with the question of quality itself, which does not necessarily improve with technology.

One reader recalled an article that the historian Barbara Tuchman wrote about the subject in 1980. Tuchman began, "A question raised by our culture of the last two or three decades is whether quality in product and effort has become a vanishing element of current civilization." She noted that her essay on the subject was not founded on "documentary or other hard evidence according to usual historical method." Instead, it represented her personal reflections backed by a half century of experience on a "pervasive problem." Anticipating pushback against what she was about to say, Tuchman prefaced her remarks by noting that

> quality, as I understand it, means investment of the best skill and effort possible to produce the finest and most admirable result possible. Its presence or absence in some degree characterizes every man-made object, service, skilled or unskilled labor—laying bricks, painting a picture, ironing shirts, practicing medicine, shoemaking, scholarship, writing a book. You do it well or you do it half-well. Materials are sound and durable or they are sleazy; method is painstaking or whatever is easiest. Quality is achieving or reaching for the highest standard as against being satisfied with the sloppy or fraudulent. It is honesty of purpose as against catering to cheap or sensational sentiment. It does not allow compromise with the second-rate.

Tuchman went on to acknowledge that "quality is undeniably, though not necessarily, related to class, not in its nature but in circumstances." She reflected on the age of the princely patron, whose wealth and power enabled him to commission the highest-quality buildings and works of art for his self-glorification, and acknowledged that we now live in "the age of the masses." To characterize this age she quoted Alexis de Tocqueville, whose *Democracy in America* was published in 1835: "When only the wealthy had watches they were very good ones; few are now made that are worth much but everyone has one in his pocket." Between the wealthy individual patron and the masses historically came the city, the state, and other beneficent institutions. They brought forth quality public works such as temples and theaters in Greece, the Coliseum and the Pantheon in Rome, public parks in London, and Gothic cathedrals throughout Europe. Whether manifestations of quality in design and execution are being created today and will be admired generations hence remains to be seen. One thing appeared clear to Tuchman, however: that our concept of quality had changed and our ability to achieve it had become threatened.

Tuchman saw the decline in appreciation for quality to have reached even "into the richer ranks, where purchasing power has outdistanced cultivated judgment." She saw the phenomenon of people of means buying and wearing designer accessories emblazoned with the designer's logo or monogram as actions that did not risk individual judgment. They paid for the stamp of approval of someone else, but in the act they were "merely proclaiming that they lack reliable taste of their own." The same could be said of people buying houses designed by trendy architects and developers or renovated by sweet-talking designers, rather than having the confidence and judgment to choose to live in a house that is thoughtfully laid out, well built with excellent materials, and practically outfitted with comfortable furniture. Tuchman saw the new egalitarianism of her time preferring "to make the whole question of quality vanish by adopting a flat philosophy of the equality of everything. No fact or event is of greater or less value than any other; no person or thing is superior or inferior to any other. Any reference to quality is instantly castigated as elitism . . ." While she admitted that "elitism is the equivalent of quality," quality does not have to be equated with money and power. As Tuchman noted in her concluding paragraph, "the urge for the best is an element of humankind as inherent as the heartbeat."

That may be true, but urges are not accomplishments. As good as he may think he is, the seasoned carpenter who cannot or does not measure a board accurately, saw it off cleanly, align it correctly, nail it squarely, or finish it neatly may be no better than a green apprentice. Too much of what I see may have been done by good-intentioned workers, but intentions are no better than urges for assuring a quality product.

The economist and *New York Times* columnist Paul Krugman, reacting to my op-ed essay in his blog *The Conscience of a Liberal*, was not sure of my assertion about the decline of good construction but did believe that "there has been a shocking and inexcusable decline in public investment at a time when we should be doing far more investment," especially in light of the high unemployment rate among construction workers—whether good, bad, or indifferent—and record low interest rates for public borrowing. In other words, the economy at the time was "awash in excess labor and capital." Was a form of Gresham's law at play, in which bad labor and materials were driving out the good? Were the better workers—the experienced ones capable of doing quality work—not willing to labor for less just to keep their skills honed and to stay in the market? Would quality, over time, not prevail over shoddiness the way good is supposed to over evil? Is there a historical cycle in which quality rises and falls like the tides, only on a much longer time scale?

Krugman admitted that "almost everyone has the sense that we used to build things better," noting that supporting anecdotes were easy to come by. He even provided a couple of examples from his own personal experience. He recalled having seen a banner on a condo construction site on the Upper West Side of Manhattan that advertised "Twenty-First Century Pre-War Living," which he translated for non–New Yorkers to mean thick walls and high ceilings, something rarely found in late twentieth-, let alone early twenty-first-century condominiums. He also confessed that the first place he lived in when taking up a position at Princeton University was a "brand-new McMansion." The huge house had gigantic rooms, but it "was falling apart from day one."

It was not too long after Krugman's blog that an article in the *New York Times* provided further anecdotal evidence of shoddy new construction, this time in condominiums in Brooklyn, where there was a building boom. The examples were located at the foot of Park Slope, not far from where I grew up. In one case, within three years of a building's completion, concrete began to flake from the façade and balconies began to crack so badly that they could not be used. The city ordered the erection of scaffolding and a shed over the sidewalk to prevent pieces of concrete from falling on passersby. Needless to say, the new owners of the condos were not pleased. Nor were residents of a nearby building that had recently been converted to condominiums: water poured into their apartments because the drainage system had not been properly connected to the sewer line. People caught up in such situations felt cheated and trapped.

Krugman worried about biased samples, thinking that maybe our supporting cases for the old being better than the new were a manifestation of a principle of survival of the fittest. Old-time buildings that were poorly made just haven't survived. Anecdotal evidence also supports this, for a casual drive along any old rural road through depleted farmland will reveal collapsing barns and farmhouses that do not look like they will survive the next heavy snow- or windstorm. But the decrepit properties are invariably unpainted, overgrown with weeds, and otherwise unmaintained. Krugman did not have to drive out of town for an example. He recalled the time when he was a graduate student and assistant professor living in Boston-area triple-decker apartment buildings whose floors were so far out of level that a rope had to be tied around a refrigerator to keep its door from being swung open by gravity.

An interesting example in support of Krugman's thesis about biased samples can be found in the history of Indiana. At the time of the state's admission to the union in 1816, legislative business was conducted at Corydon, which is located in the extreme southern part of the relatively long and narrow state. By 1825 a more centrally located seat of government was established at Indianapolis. Alexander Ralston, who had apprenticed and worked under Pierre Charles L'Enfant, who laid out Washington, D.C., was commissioned to plan the permanent Indiana state capital. Within a mile-square plat, Ralston laid out a grid of nine east–west and nine north–south streets, each a generous ninety feet wide. The central four blocks were designated Governor's Square, and in the middle of it he designed a circle around a raised knoll—a large common intended as the site of the governor's mansion.

Two years after the legislature first met in the new capital, the assembly appropriated $4,000 to build a house for the governor. Its layout was appropriately square, and a large room was located at each corner. Between the rooms, wide hallways ran the length of the house in each direction. The two-story yellow-brick structure was imposing, but its interior, although spacious and symmetrical, was not well planned for comfortable family living: the rooms were drafty, there was no convenient kitchen, and the basement was dark and damp. The house was completed in 1827, at an actual cost of close to $6,500.

James Brown Ray, an Independent who served two and a half terms, was the first governor able to use the building. However, when his wife saw the property, she refused to live in a house with such an exposed yard; neither the Rays nor any subsequent governor and his family occupied the putative mansion. Rather, its tenants included at various times state judges, the state library, the state bank, and a kindergarten; the house also was the venue for charity events, celebrations, receptions, balls, lectures, meetings, and fairs. But within about two decades of its completion, the building began to deteriorate and fell out of favor for festive gatherings. The city's newspaper noted with sadness the condition into which the structure and its grounds had declined, describing it as "shabby" and an "eyesore": "The grass was trodden, the fence rickety, the trees ragged, the ground covered with ugly shrubbery, sticks, stones, old shoes . . ." The newspaper wished the "dilapidated" building and grounds could be replaced by something more in keeping with the aspirations of a proud city. In 1857 the thirty-year-old structure was sold at auction for $665; it was demolished and salvageable materials were used in the construction of a new hotel nearby. The vacant lot "became a dirty treeless expanse where cows grazed and pigs explored the rubbish of the damp basement." After the Civil War, the land was developed into a park—Circle Park. After some time, a monument to Civil War veterans was proposed to be erected there. Eventually, a design competition resulted in the State Soldiers' and Sailors' Monument that was completed in 1901. It towers almost 285 feet above street level and dominates the circle, appropriately known now as Monument Square.

Neglecting real estate and letting it deteriorate is nothing new. Apartment dwellers have known this for some time, and the conflict between landlord and tenant is classic. In the television sitcom series *The Big Bang Theory*, the out-of-service elevator—marked with yellow caution tape and around which the stairway winds like a square-cornered helix—was a permanent part of the set even as the show was going into its eighth season. Neither the characters nor regular viewers seemed to expect the elevator ever to be fixed. Apartment dwellers Sheldon and Leonard—occupants of Apartment 4A—and Penny—across the hall in 4B—resigned themselves, as did their frequent guests Howard, Raj, Amy, and Bernadette, to walking up and down multiple flights of stairs, sometimes hauling laundry and other bulky and heavy burdens, with the scenes in which they did so suggesting a Sisyphean journey around an endless stairway of the kind that the graphic artist M. C. Escher made famous.

Apartment buildings in large and small college towns can be run by landlords who see their tenants as people who will soon move on and so whose requests and complaints can easily be ignored. Sometimes the government can appear to act like a slumlord, neglecting the maintenance and repair of subsidized public housing developments—what in many larger cities are known as "the projects." This neglect can be exacerbated when money is tight generally, for the projects have traditionally been home to people who do not have much political savvy or influence and therefore whose needs could rather easily be ignored by bureaucrats. For a similar reason streets and sidewalks in poorer neighborhoods are often the most neglectfully maintained. In a large city like New York, where in 2014 more than four hundred thousand people lived in apartment buildings owned by the city's housing authority, the situation worsened when federal subsidies were cut and the backlog of repairs and upgrades amounted to $18 billion. Under $4 billion of this work was expected to be done over the subsequent four years. This came at a time when the poor were also facing high unemployment.

A thoughtful opinion piece in the *New York Times* by Jayne Merkel put forth a proposal that could ameliorate both problems. Its author, a contributing editor of *Architectural Record*, suggested that the New York City Housing Authority "train tenants to do basic maintenance," such as painting, plastering, and fixing leaks, thereby providing job training while realizing improved apartments. Merkel pointed out that the idea is not novel, for "in many parts of Europe, public housing residents are required to take responsibility for some maintenance." It is an idea akin to that which allowed, in many parts of the young United States, citizens to work off their taxes by working on the road. As with the roadwork they produced, however, allowance would have to be made for the quality of maintenance done by trainees. But everyone has to learn by doing.

In his blog about infrastructure spending, Krugman noted that the period after the Civil War was described by contemporaries as the Age of Shoddy, and reminded us that the noun "shoddy" originally referred to an inferior form of yarn or fabric made from old woolen rags. That meaning of the word dates from the 1830s, about the time that balloon-frame building construction was introduced, which, when compared to the traditional post-and-beam method, might have been considered— and in many cases may have been—shoddy indeed. Many of the early structures framed in scantlings—the two-by-fours of the day—would likely have suffered in workmanship, because among the appeals of the new method was that the pieces of lumber forming the building frame were connected together only with nails, rather than being assembled with solid peg-locked mortise-and-tenon joints of traditional heavy-timber construction. While practiced carpenters may have been able to build a sturdy balloon-frame house, a typical amateur would not likely be able to do so perfectly the first time or two. Hence we would expect many of those early attempts to have long vanished from the face of the earth. But over the next century, in the hands of practiced carpenters, whether amateur or professional, the balloon-frame structural system became a solid one indeed.

The disappearance of well-built and -maintained houses from the late nineteenth and early twentieth centuries is not so much proof of poorer quality materials and craftsmanship as it is of changes in taste, fashion, and aspiration. Just look at the Detroit neighborhoods in which large old houses stand among vacant lots. When the city and the neighborhood were thriving, contemporary houses built cheek by jowl to more or less the same standards must all have had about the same chance to grow old gracefully. That some did not was likely a result of neglect rather than fundamental structural difference.

Sometimes even the best-built of houses perished with the poorly made. In the aftermath of the Great Chicago Fire of 1871—in which streets paved in wood blocks soaked in creosote fed the conflagration rather than served as firebreaks the way stone streets elsewhere did—about one hundred thousand residents of the city found themselves homeless. To help remedy the sudden housing shortage, home-building kits were made available to the victims. There were two sizes of these small so-called fire-relief or shelter cottages: a 12-by-16-foot model intended for families of three and fewer, and a 16-by-20-foot version for larger families. The 192-square-foot style was available for $75 and the 320-square-foot one for $100; the Chicago Relief and Aid Society often gave them away for free. All the receiving family had to do was provide the land and labor to assemble the precut lumber into a cottage. Five weeks after the fire, about 5,200 of the houses had been erected; six months later another 3,000 had gone up. But since wooden structures had fed the great fire, in time the erection of more wooden cottages was halted by reconsidered building codes.

Today, no one knows for sure how many of the shelter cottages remain. As with the tract houses in Levittown, New York, the development that became synonymous with the postwar American cookie-cutter suburb, so many of the fire-relief cottages underwent expansion, remodeling, and other alterations—not to mention demolition to make way for larger structures—that it is not easy to locate one in original shape. But two are believed to have survived in near-mint condition, and at least one has become a point of interest on Chicago city tours. These two may be considered survivor-bias data points, but the fact that they have lasted for almost 150 years suggests at least that materials of sufficient quality were likely also used in the non-survivors of this standardized type. And the workmanship could have been as good or better. In building, at least, survivor bias can tell us more about mores than about quality, for the shelter cottages that no longer exist may have succumbed to the need or desire for more space, a grander and more fashionable façade, or a change in style rather than of substance.

It is not so much that old construction was cheap and unsubstantial. As building practices evolve, builders gain by experience knowledge of not only what works well and what works poorly but also what is profitable and what is not. Florida experienced some devastating hurricanes in the 1960s, when houses lost their shingles, their roofs, and their walls. Failure analysis revealed what weaknesses of construction led to the damage and destruction, and more stringent building codes and construction practices were instituted. Thus, houses built during the time of heightened awareness were solid and robust. However, for decades afterward no major hurricanes hit the area, and so fear of their destructive power receded. This was manifested in more casual construction practices that, when they did at all, adhered to inadequate building codes. Fewer nails or weaker staples were used to install shingles, insufficiently anchored framing was employed, and generally more lax attitudes on the part of builders, inspectors, and buyers alike prevailed. Only when Hurricane Andrew struck in 1992, causing a record $26 billion in damage, did the exposed weaknesses in newer homes resensitize everyone to what can happen.

We can expect that building generally follows such a reactive cycle, whose period lengthens and shrinks with that of the short-term memory that residents and bureaucrats alike seem to favor over the long term. By relaxing our fears and standards as memories of how fierce the wind can be recede into the past, we construct houses and other buildings that are incapable of standing up to the next big storm. This is human nature, and even the most conscientious of city engineers, town managers, and code-writing bodies are susceptible to such fallibility. They relent when they hear over and over again that the last big storm was an outlier on the plot of prediction and will not happen again in a thousand years. Consequently, building codes and their enforcement slacken.

Structures that were built with integrity to begin with also can be compromised when they need maintenance and repairs. When the University of Southern Maine announced that it was going to install vinyl siding (and modern windows) on an almost two-hundred-year-old building that had been faced with white pine, preservationists objected. The historic structure had at different times been used as a meetinghouse, the town hall, and a chapel, and was now a campus art gallery. The preservationists argued that historical integrity called for using white pine, with a three-inch reveal, fearing that otherwise the 1821 Greek Revival structure would be removed from the National Register of Historic Places and so be more vulnerable to being allowed to fall into neglect—and eventually to the bulldozer. The university had budgeted $320,000 for renovation work; using pine would have added about 10 percent to the cost, and white pine, which is not a durable wood, would not last as long as vinyl, because the white pine available today is not equal in quality to that used in the nineteenth century. Indeed, newer pine would have to be repainted every five years or so to preserve it. Reluctantly, because the university had been facing budget problems, it relented and agreed to use wood instead of vinyl (and to restore the old windows rather than replace them with new ones). The move toward a seemingly inexorable decline was only narrowly averted.

At the end of his blog/post, Krugman admitted that "maybe we really are building worse" today, but he would like better evidence to support the thesis. Anecdotal evidence from owners and the structures themselves does support the thesis that surviving American homes from the earlier and middle part of the twentieth century were made more substantially than those built at the end of the century. Whether the sample available to us is biased because the shoddy ones fell into decline and passed into oblivion does not invalidate the claim. To say that surviving older homes were made better than newer ones is not to say that there were no shoddy homes built in an age of quality. Those that were simply did not last or survive. The important point is that the paragons of quality from the 1920s or the 1950s show what was achievable in those times—and should be now. And there are blocks and blocks of such homes standing today in neighborhoods like St. Albans, located in the New York City borough of Queens, and Oak Park, Illinois, a village just west of Chicago proper where Frank Lloyd Wright spent the first two decades of his architectural career. Wright left behind numerous examples of his work in Oak Park, which have been preserved for their historic significance, even though many a Wright building has proven to be a homeowner's nightmare. Quality does not derive from a designer label but from a solid design and its execution. Wright's Fallingwater, for example, created as a weekend retreat in the woods of southwestern Pennsylvania, a ninety-minute drive from Pittsburgh, is infamous for the necessity of having its dramatically cantilevered concrete balconies reinforced because of their poor initial structural design. Of the approximately 415 houses and other structures designed by Wright between 1886 and 1959, some fifty-six (or 13 percent) no longer stand, not because they were necessarily of inferior Wright quality, but because they succumbed to fire, storms, or demolition. It is not always for lack of quality—of the design, the materials, or the construction—that buildings do not survive. Nor is it solely because of quality that the new displaces the old in domestic or public infrastructure.

We need to take a holistic view of infrastructure to understand why something survives and something else, even though almost identical, does not. In the case of individual pieces of architecture, the principle of survival of the fittest does not necessarily apply, since what is fittest depends upon when and where and by whom the field is being judged. So it is with infrastructure generally. A bridge perfectly fit structurally may not live out its design life because it is not fit functionally. Thus, many a highway bridge has been replaced before it absolutely had to be for structural reasons because the roadway it carried—or the one over which it crossed—had to be widened from two lanes to three, or from three to four. Whether the new bridge will be as well built as the one it replaced will depend on many things, including but not limited to the nature of the design, the quality of the materials used, the care with which they are assembled, and the rigorousness of inspections of the work. And more than any of these factors, whether the new bridge will last as long as the old may depend more on when the road will next be widened.

14
—

About the Same

GOOD ENOUGH BRIDGES AND BAD ENOUGH TOOLS

BRIDGE FAILURES FOLLOW A cyclic pattern, occurring as they do after a prolonged period of unremarkable service. As I have written elsewhere, this period of perceived quality known as success lulls engineers into thinking they have solved all the relevant problems with bridge design and construction, and so they relax their guard, their standards, and their vigilance, meaning that they design less conservatively, build with less attention to detail, inspect more casually, and generally lower the quality of bridges being built. In the meantime, they also tend to push the limits of their technology, making longer spans and more daring new structures. The duration of success is typically remarkably close to three decades—about the extent of a professional generation—and by then complacency has become so common that any warning signs of impending disaster there might be are neither noticed nor heeded. When a major bridge failure does occur—like the destruction of the Tacoma Narrows Bridge in the wind in 1940 or the spontaneous collapse of the Minneapolis Interstate 35W bridge in 2007—it is a clear wakeup call that complacency had indeed set in. Subsequent to the failure, engineers—with the public and politicians looking over their shoulders—gain a new sense of urgency to design and oversee construction and maintenance with less carelessness and bravado and more quality and care. And so the cycle repeats.

Even given such a cycle, it is a fact that bridge failures are simply not very common. One major failure every thirty years is certainly no indictment that engineers and builders are without some considerable understanding of the workings of the daring structures that carry automobile, truck, and rail traffic across wide rivers and valleys. But the accusation of survivor bias with regard to praising the quality of older homes might make us wonder if such a phenomenon is also at play with regard to bridges. Were older bridges made better than newer ones? Any suggestion that they were is certainly counterintuitive. Nevertheless, we do celebrate ancient Roman aqueducts still standing in France and Spain; the countless centuries-old stone bridges still in service throughout Italy and the rest of the world; the first iron bridge, which has been standing since 1779 at Coalbrookdale, England; and, of course, the Brooklyn Bridge, which since 1883 has been faithfully serving New York City commuters.

At the same time, we see highway bridges that have served our interstates for mere decades being torn down after being replaced with newer, wider models. The multi-span Tappan Zee Bridge, when barely a half century old, was being replaced with a grand new crossing because the old steel cantilever was not considered up to current structural or functional standards or worth maintaining. And as we have seen, the east span of the San Francisco–Oakland Bay Bridge, which early in its sixth decade was damaged by an earthquake, was entirely replaced at a cost in excess of $6 billion. Did they or did they not really build them better in the old days?

The Eads Bridge across the Mississippi River at St. Louis is an iconic structure that in 2014 reached its 140th anniversary. The bridge had not been without trouble: its construction claimed the lives of fifteen men who worked in the pneumatic caissons driving the river piers down to bedrock; the bridge's daring arches made of the new material steel were extremely difficult to complete; the bridge company went bankrupt within a year of the span being opened to traffic; part of the bridge's east approach masonry arcade was destroyed by a tornado in 1896; railroads stopped using the bridge in 1974; and in 1991 its upper roadway was closed to motor vehicle traffic for major rebuilding. The structure was reopened to automobiles in 2003, a decade after its lower deck began to be used by Metrolink, the light-rail system that serves the St. Louis metropolitan area. Yet in spite of its vicissitudes, the Eads Bridge is a symbol of American innovation and resolve, and a clear success in its innovative use of the steel arch. It holds its own, even if as a low-profile architectural and engineering landmark in the shadow of the boldly soaring Gateway Arch.

The Eads Bridge and the associated surface and tunnel infrastructure that integrated the river crossing into the city's existing street and railway network were clearly built to last, and they carried a correspondingly serious $10 million price tag. There were other river crossings proposed for St. Louis in the mid-nineteenth century, and had one of them been built instead of the Eads, it is not likely that it would still be standing in its original form the way the Eads essentially is. The most likely alternative was one promoted by Lucius Boomer, a bridge

The Eads Bridge was under construction from 1868 to 1874. Its arches were made of steel, which was stronger but more expensive than iron, the more common structural material of the time. Unlike most of its contemporaries, the Eads Bridge was not built in a shoddy manner, nor was it expected to be replaced before too long. Indeed, it remains a part of the infrastructure of St. Louis and southern Illinois.

financier, and the engineer Simeon S. Post, whose eponymous truss was to be the main structural element used in their bridge.

Although the railroads clearly needed bridges to carry their tracks across rivers and valleys, they did not necessarily have the finances to do so at will. According to historian of technology John K. Brown, the iron truss bridges of the mid-nineteenth century were generally made only "good enough," the bridge-building companies giving the railroads just what they wanted—and very little more. But the railroads were investing in more than rails and bridges; they were also developing heavier and heavier rolling stock. This meant that in as short a period as a decade or so bridges that had once been good enough no longer were adequate to carry the increased loads. Older bridges had to be strengthened or torn down and replaced before they collapsed under a train. Thus, the fate of a bridge built in the 1860s might be to last only into the 1880s, and the replacement only into the early twentieth century. The Eads Bridge, with its use of the stronger material steel and with its robust design, was an exception. Were it not for James Buchanan Eads's seeking and securing investment capital outside the railroad business, his bridge, which came at a price at least twice what a good enough one would have carried, might never have been completed.

Given the way bridges were financed, designed, and built in that earlier era, there obviously can be a clear danger of employing survivor bias in trying to assess the quality of historic bridge-building technology only through its survivors. But in the case of railroad bridges, at least, no one wished to build structures that were not good enough to serve at the time. After all, no railroad would have wished to see a bridge collapse under a speeding train. Lives likely would be lost, significant cleanup and repair costs would be incurred; and since a bridge collapse would be big news, the publicity would not be good for business. A railroad may have wanted its bridges only to be good enough, but that goal could still demand a quality product, relatively speaking. Looked at from this point of view, concern over survivor bias does not apply. A bridge torn down or replaced by design because it had reached the end of its intended useful life should not be judged against examples like the Eads, which might have been considered a contemporary extravagance. And a bridge built within its budget should not necessarily be considered a substandard structure in its own context.

Today, highway bridges can face ever increasing loadings analogous to those that the railroad bridges of the nineteenth century and road bridges of the early twentieth did. How heavy a truck is allowed to use a highway or a bridge is decided not by the engineer designing the structure but by legislators yielding to pressure from lobbyists, such as those representing the trucking industry. When a bridge is on the drawing board, engineers size its beams and girders to carry a load imposed by trucks permitted by then-current laws, plus more via a factor of safety. Those laws might limit vehicles using a certain class of highway to a gross weight of, say, 80,000 pounds, with that weight typically distributed over five axles. The road and its bridges will have been designed to these specifications, but future legislators may increase the limit to, say, 110,000-pound trucks, with no change in road or bridge structures. Michigan has allowed trucks weighing up to 164,000 pounds, typically distributed over eleven axles, to use its highways without special permits. Trucks with gross weights as high as 1.5 million pounds have been able to pay nominal fees to obtain special permits to travel over Michigan roads. Factors of safety employed by the designers clearly do enable heavier loads to be carried, but if that were done routinely, highways and bridges would age prematurely. In 2014, proposed legislation to cut the weight limit in half in order to reduce damage to roads was rejected decisively by the Michigan state senate.

Because of reserve strength built into them, today's roads and bridges are able to support greater weights than they are nominally designed for, but they will do so at a cost. Allowing heavier trucks to use a road and its bridges will cause them to deteriorate at a faster rate than anticipated by engineers, and so whatever long-range planning a state highway department may have done will have to be revised, for the road surface will develop potholes faster and the bridge structure cracks earlier than predicted. The upshot of this will be, of course, that repaving of the roadway and strengthening or replacement of the bridge will have to be done earlier than planned. This wreaks havoc with long-range budgeting, and all the advances in careful, thoughtful, and rational transportation planning are essentially nullified. Dealing with infrastructure is thrown back to earlier times, when repair, maintenance, and replacement were done reactively rather than proactively. Such predicaments can only be done away with if the different branches of government work in a cooperative rather than an autonomous way.

"GOOD ENOUGH" WAS THE norm in more than just bridge building in early America. Historian of technology Eugene Ferguson, writing in the late 1970s about the "American-ness of American Technology," observed that this was due not only to a scarcity of capital but also to, in the historian John Brown's words, "the self-fulfilling perception that rapid innovation would quickly render current designs obsolete." There was a "national bias for cheap initial construction."

According to Ferguson, this tradition began in the colonial period, when a shortage of labor led to a drive toward mechanization. Since virtually no infrastructure existed in the early colonies, everything had to be built at a pace much faster than that prevailing in Europe. For a million households to have been established in the United States by 1800, and another million by 1825, houses had to be built at the rate of one hundred a day. It was no coincidence that American ingenuity developed the balloon-frame structure, which could be erected so much more quickly and with less skilled labor than a traditional one. Expedience and impermanence necessarily characterized much of what was built.

This was, of course, in sharp distinction from the European craft and guild tradition. As late as the early twentieth century, in a comparative study of industrial life in England, Germany, and America, the English encyclopedist Arthur Shadwell—quoted by Ferguson—could characterize the American way with the phrase "Let it go at that," for to his eye this seemed "to be written all over the face of the land. You see it in the wretchedly laid railway and tramway tracks, in swaying telegraph poles and sagging wires, . . . in streets unscavenged, in rubbishy cutlery . . . in scamped and hurried work everywhere. There seems to be a disdain of thorough workmanship and finish in detail." Shadwell also found it "surely remarkable that so little first-class work of any kind is produced in the United States, with all its wealth, population, intelligence and intellectual keenness." He did acknowledge, however, that "there are exceptions," that quality did exist. Examples of this—in the form of elegant town houses and office buildings—certainly have survived and provide the biased sample that continues the rosy impression of the good old days.

Ferguson noted that democracy was seen as the prime force to "spur American technology to provide good enough to satisfy as far as possible the wants of everyone." Perhaps not surprisingly in this context he cited Tocqueville as arguing "that the tendency of a democracy was to encourage the production of more goods for everyone; at the same time he thought it inevitable that the producer would be driven to making an imperfect product and the consumer forced to content himself with imperfection." This can be the downside of seeking more things for more people, as Ferguson reminded us in quoting the European social commentator Richard Müller-Freienfels: "Quantity, in America, is not a fact, as it is with us; it is a value." Today, however, the American producer and consumer are not just interacting with each other. World trade gone amok has brought us products so imperfect that consumers cannot so easily be content with them.

During a particularly hot and humid summer in North Carolina, several doors in my house had swelled to the point where they did not move freely in their frames. The problem was obvious, and one weekend I determined to plane, sand, or shave down the sticking corners. To accomplish this seemingly simple task, I shopped for a tool sold under the name Surform, which is a cross between a plane and a rasp. I still had the one I had bought decades earlier, but its blade had gotten dulled from so much use then and corroded a bit from so much disuse since. A visit to a home improvement store revealed an unexpected surfeit of Surform models—all now made in China—and I chose the one that most reminded me of my trusty original. At home, I began to work on the doors, but the tool's blade broke after just a few hard swipes at the wood. It appeared that the metal had been embrittled in the manufacturing process and so was prone to fracture at a weld near where the blade was attached to the frame. The broken tool was replaced by the store, but before I could finish my task the second Surform failed in the same way. It, too, was returned and this time exchanged for a slightly smaller model, in the hope that the manufacturing fault would not have affected its blade design. I used this tool with extra tender care, wanting only to finish the job of freeing the doors in their frames. I did accomplish the task, but with no love developed for the tool.

So many carpenters' tools—along with clothes, electronic devices, and building materials—are now made in China that it is difficult to find much choice among those that are not, if any indeed can be found in a nearby store. Without quality tools, it is very difficult to build quality structures. For example, I have two steel measuring tapes of the kind that come coiled up in a hand-size case. One of these tapes, ten feet long when fully deployed, is labeled a "professional" model and is graduated in thirty-seconds of an inch (and tenths of a centimeter, or millimeters). The other, a thirty-footer, is graduated only in sixteenths of an inch. On the face of it, the so-called professional model can be used to measure more accurately and could do so outside the United States, an anomalous holdout against the metric system.

When I used both tapes to measure a small wooden box with sharp square sides, the first tape told me that its outside dimension was $4^{23}/_{32}$ inches, whereas the second tape said $4\frac{3}{4}$ inches. The difference of $^1/_{32}$ of an inch appeared to be due to the movable hook that forms the zero end of the tape, which is designed with a bit of play to compensate for its thickness when measuring inside versus outside dimensions of something like a door or window frame, or little wooden box. As long as I consistently use one of the tapes, the absolute measurement will not matter, but if I were to measure with one tape what size opening I want to fill with a piece of wood and with the other tape mark the piece I need to saw off a longer length of lumber, I could end up with something too long or too short by $^1/_{32}$ of an inch. (The same thing would happen if I were to follow dimensioned plans and use an inaccurate measuring device.) This is the thickness of the piece of cardboard backing of the legal pad on my desk, and could make for either a too-tight fit or leave a noticeable gap in my carpentry work. Either way, it would be a blemish on the work and would not say "quality." Perhaps not even "good enough."

The first tape—the self-declared professional one marked off in thirty-seconds of an inch—itself does not look professional or exude an air of quality. Its plastic case is of a curious blue-green color, and the visible heads of the screws that fasten the back to the front and hold the tape in place are corroded. It bears no brand name, but it does have a small shiny label affixed to the back that reads, "Made in China." My other tape carries the name of a well-known American tool manufacturer; I assume that it is made in the U.S.A., but I cannot find on it any confirmation of that. It has a substantial chrome-like case, albeit also plastic, and contains a wider and longer tape that is very much heavier. The screws fixing the back to the front have been exposed to the same climate as the other tape, but these are not rusted. Overall, the difference is striking, but the more coarsely marked tape looks, feels, and functions as the more professional model, even though not labeled as such. The good carpenter, in a manner analogous to an experienced engineer reading a slide-rule's logarithmic scale, will easily be able to interpolate by eye between linearly spaced $1/16$-inch marks to measure accurately even to $1/64$ of an inch, something actually less easily done on the crowded scale of the Chinese model. Quality tools and the quality workmanship they make possible should be judged more by the hand of the user than by the eye of the beholder.

Ironically, the Surforms whose blades fractured after only a few swipes at the edge of a sticking wooden door were sold under the same brand name as the measuring tape that I found to be superior. But why would a manufacturer risk its reputation by selling inferior tools? No doubt the Chinese-made Surforms were less expensive to manufacture and so could be sold for a more competitive price in America. And presumably the U.S. company tested the imported tools before associating its name with them. But did it continue to monitor their quality? Weekend carpenters will appreciate being able to save a few bucks on a tool they may use only now and then, but they do not expect it to break on the first use. How long carpenters and do-it-yourselfers will tolerate inferior imported tools remains to be seen, but the predominance of poorly made Chinese products is likely to be part of a cyclic movement between quality and shoddy that will play out and give rise to improved imported tools or a resurgence of superior American-made tools, even if they are more highly priced.

In this kind of dynamic, hope for a resurgence of domestic manufacturing resides. And it should not be expected only in hand tools. The disappointing quality of foreign-made steel imported for projects like the Bay Bridge gives rise to renewed interest in a revitalized domestic steel industry. And because this industry will know that it owes its existence to the unacceptably poor steel from abroad, quality should be an expectation and a priority. As long as the new industry can continue to produce a quality product, it will thrive. If the revitalized industry becomes complacent, however, it may allow poor steel to drive out the good and encourage the development of a vicious cycle.

15
—

Lay in Leaves

TRIAGE, BUDGETS, AND CHOICES, AGAIN

To PAVE OR NOT to pave? That is the question. We live among streets that are in varying states of disintegration, but we also live in an economy of limited resources. How do we triage the cracked, the potholed, the disintegrating? What is the optimal strategy for maintaining roads within a tight budget?

One approach is termed "worst-first." This practice of repairing obviously poor roads before tackling those that are only beginning to show signs of deterioration will generally not encounter much opposition from the citizens of a community. Everyone can clearly see how badly the chosen road has deteriorated, and so who can be against its immediate repair? However, unless the repairs to the road are done to the extent that they need to be, it may be in the same or worse shape in a couple of years. For example, if the principal problem was that the subsurface was not draining properly and so allowing potholes to develop, just resurfacing the road will not eliminate the root cause of the problem. But to redo the road down to its foundation would obviously cost a lot more than just resurfacing. Choosing the more expensive route would leave less money for fixing other roads in the jurisdiction, and allowing them to go another season or two without attention may mean more to repair them in the end.

According to construction industry writer Tom Kuennen, a road that has held up very well for years can suddenly reach a point where it deteriorates so quickly that it cannot be saved by simple repairs. Experience has shown that "spending $1 on pavement preservation before that point eliminates or delays spending $6 to $10 on future rehabilitation or reconstruction costs." The preservation strategy might mean spending on "maintenance of a pavement even when there is nothing apparently wrong with it." This, of course, could be a tough sell where roads in obviously poorer condition are routinely neglected. Still, in 2005, the director of the Office of Asset Management of the Federal Highway Administration declared his support for the concept, which he described as "a paradigm shift from worst-first to optimal timing." At the same time, he acknowledged the need for a focus on "convincing the public, and selecting the right treatment for the right pavement at the right time." Ah, there's the rub.

Convincing the public may not be easy when the options and details are as numerous and technical as those outlined by Kuennen in his article in the trade journal *Asphalt*. The importance of some declared good practices are self-evident, like the value of "sweeping and sealing the joints and cracks" in an old pavement before applying a new surface to it. The cautionary advice of Caltrans, which he quotes, also makes sense: "It is critical that all necessary preparation work such as crack filling, pothole repair, patching, leveling, and dig-outs be done prior to surface treatments being placed." Since Kuennen was writing for the trade, it is understandable that he eased into jargon like "dig-outs" and, elsewhere, terms like "fog seals," "sand seals," "slurry seals," and "chip seals." The details are important, of course, but if local road departments wish to convince the public of the wisdom of their recommendations without getting deep into the weeds, they will have to use more down-to-earth language.

In an article laden with shoptalk, Kuennen gives a simple definition of an asphalt pavement's greatest malady: "Potholes are bowl-shaped holes of various sizes which are associated with pavement fatigue and poor drainage." He advises that potholes can be minimized by "keeping water out of the base material," thereby keeping the pavement support from being weakened and eliminating one of the contributors to frost heave and cracking—the latter often occurring in a mosaic pattern referred to as "alligator" or "crocodile" cracking for its resemblance to the animals' skins. The near-rectangular, tile-like arrangement of material bounded by the cracks begins to flake or pop out, producing the beginnings of a pothole of irregular shape. Sometimes the edge of the pothole approximates in geometry a polygonal circle, but because the initial cracking pattern is composed of skewed square-tending-to-rectangular cells, potholes with oblong shapes also develop.

However formed and shaped, everyone knows that a zero-pothole goal is not very practical, and so Kuennen cites a long-term study on patching the holes. Not surprisingly, it found that the single most important factor in getting a good patch is to use quality materials. This proved more important even than the technique used in applying the material to the damaged pavement. Thus the so-called throw-and-roll method, in which an adequate amount of the patching stuff is simply thrown into the hole and then compacted by being rolled over with the tires of the maintenance truck, was found to be as effective as what is known as a semipermanent repair, in which the edges of the irregularly shaped hole are first cut square and the patching material compacted with a vibratory tool. Obviously, the former method is simpler and quicker, and consequently costs less per pothole to implement. As the terrible winter of 2015 wound down, the New Jersey Department of Transportation even introduced an improvement on the old throw-and-roll method. It deployed a fleet of thirteen "Pothole Killer" trucks, each operated by a single worker who did not have to leave the driver's seat to fill a pothole. By means of remote-controlled tools at the end of a hydraulically operated arm, the hole was first cleared of loose debris with a blast of compressed air, then sprayed with a tacky substance and finally filled with patching material. After the Pothole Killer moved on to its next job, the tires of highway traffic took care of the compacting. With no maintenance men getting in and out of the truck, the procedure promised to be safe and efficient. Approximately three hundred thousand potholes were expected to be filled by the fleet in the course of a few months.

For "severely distressed surfaces" whose repair is beyond the capabilities of a Pothole Killer, Kuennen recommends, rather than just filling the depressions, applying to the entire road surface a 0.5- to 1.5-inch-thick overlay of hot-mixed asphalt, which is delivered to the paving site with the stone- or rock-chips aggregate mixed in. The high temperature of the mixture makes it flow smoothly over the road surface. However, this feature can be negated if the truck delivering the mix has to travel too far after being loaded or if the ambient temperature is too low. Hence, the overlay method has its practical limitations of being suitable for roads only within a certain distance of the asphalt plant and only during the warm season. Nevertheless, for what Kuennen describes as "only an incrementally higher expenditure" of maintenance dollars, the procedure provides, "aesthetically, the overall impression of a brand-new road, at the price of a thin overlay."

The town of Hancock, New Hampshire, has some older roads that date from the eighteenth and nineteenth centuries. Admitting that such a collection of roads presents special challenges, the town's director of public works, Kurtis Grassett, wrote a paper outlining the issues involved in making choices in maintaining paved roads and explaining concisely the pros and cons of the different options available to do so. In a sobering sentence, he wrote that "we know that all roads will fail." The question is not if but when. The complex cycle of use and abuse of a road in a northeastern state makes the answer less than obvious, but the so-called failure curve that the director describes shows that, once a road begins to deteriorate, the downward trend proceeds rather precipitously.

The trick is to invest in rehabilitation of the failing road at a time in its life cycle where the most bang for the buck is received. According to Grassett, in the span of 75 percent of a roadway's life, it will decrease 40 percent in quality, going from an excellent new road to a fair older one. Sounding like Kuennen, Grassett reminds us that neglecting to invest $1 at this stage to improve a road's condition back to good or better will incur a cost of $4 to $5 just a few years later. Of course, knowing exactly where a road is in its life cycle involves experience and judgment, and this is what directors of public works and highway engineers are paid to have. Among a typical professional's activities is to survey all a town's roads every two years or so—looking for cracking, heaving, rutting, and the like—and comparing the current state with past observations. According to Grassett, "This helps us to estimate the rate of failure and allows us to choose the maintenance technique that will give us the biggest return on investment."

Perhaps to keep the choices of maintenance procedures to a manageable number in his exposition, Grassett limited his advice to four basic paving operations, and perhaps most significantly he pointed out how long each treatment could be expected to last before the road had to be resurfaced again. He considered three sealing procedures and described the concept of shimming, whereby the base of a road is properly shaped before being sealed. The choices are summarized in the following table derived from his paper:

A COMPARISON OF ROAD MAINTENANCE PROCEDURES

Type	Procedure	Service life	Cost	Comments
Crack sealing	Clean crack; place hot asphalt liquid	3–4 years	$1–$2 per lineal foot	Works best on fairly newly paved surface
Sand seal	Apply hot liquid asphalt, cover with sand, roller over sand	4–6 years	$2 per square yard	
Chip seal	Apply hot liquid asphalt, cover with layer of 3/8-inch stone chips	6–8 years	$3.25 per square yard	Used on roads with steep grades and heavy truck traffic
Asphalt shim	Lay down thin (less than 1-inch) layer of asphalt over existing surface	Lifetime determined by sand or chip seal applied over shim	Can be included in sand or chip seal process	Restores crown to road surface; often used in combination with sand seal; done within two months of sand seal

Of course, if a road has deteriorated to the point where maintenance would not be cost-effective, the road may have to be reconstructed from the ground up, which entails a further set of choices. What the right treatment might be in a given situation depends upon a number of factors, not least of which are the subjective ones of judgment and preference. And how wisely these will be applied will depend on the knowledge, experience, and integrity of the people involved in making the recommendations and decisions about priorities. This is as true of large cities as it is of small towns. In the case of Hancock, New Hampshire, Director of Public Works Grassett informed the citizens of the town that after studying the situation, the return on investment in road maintenance was not satisfactory, and he was spelling out the options so that the town could make an informed judgment about what to do. Most likely, to get better roads, they would have to vote themselves an increase in taxes.

Since the construction and maintenance of local roads are generally the responsibility not of the state or county but of the city, town, or village in which they are located, the citizens are often asked to approve a budget, a tax increase, or a bond issue in support of new or better roads. In Maine, the small island-town of Arrowsic, with a 2010 census population of 427, has a road commission that meets monthly and reports annually. The budgeted salary for the four-member commission is $541 per year, which includes social security and Medicare taxes. In addition to having responsibility for assigning road names and house addresses, installing road signs, and coordinating the town's emergency database with that of the state, the commission oversees the resurfacing, maintenance, and snowplowing of roads other than the state highway that runs the length of the island. For the small town, "a significant paving effort" means improving a section of one of the town's roads leading off from the highway. For a paving contractor, the job is "a small project."

For the fiscal year 2014–2015, Arrowsic asked its residents to approve for general maintenance and repair of town roads, including snow removal, a budget just short of $175,000, which broke down as follows:

ARROWSIC ROAD COMMISSION PROPOSED
BUDGET, 2014–2015

General Maintenance		$2,500
Culverts		1,500
Roadway Grading/Rebuild		10,000
Asphalt and Gravel Maintenance		3,000
Road Sign Replacement		1,000
Paving		75,000
Emergency		1,500
Snowplow		80,000
	Total	$174,500

This amounted to just over $400 per citizen. The bulk of the budget ($165,000, or 95 percent) was for grading, rebuilding, paving, and snowplowing of roads. Or, looked at differently, about the same amount of money was budgeted to clear the roads of snow as was budgeted to fix roads that snowplowing no doubt contributed to damaging. Such is the nature of infrastructure: using it means wearing it out, just like everything else. Of the total amount of the Arrowsic road budget, $8,600 was expected to come from a block grant from the Maine Department of Transportation and $85,000 from town excise taxes. The remaining $80,900 was to be raised and appropriated by the town itself, mainly from property taxes. Aside from education expenses, which amount to about half a million dollars annually, the roads budget takes the largest slice of Arrowsic's budget pie.

Since Arrowsic's principal road is a state highway, the Maine Department of Transportation takes care of it, as well as the bridges on that road that carry traffic onto and off the island. As small a town as Arrowsic is, its improved roads are among the most important amenities of modern island life. To ensure that the town roads are not abused when the spring thaw and mud season make them most vulnerable to damage by heavy trucks, the road commission posts weight limits.

Typically from mid-March to early May, depending on the weather, vehicles heavier than 23,000 pounds gross weight are restricted from the posted roads. This helps keep maintenance and grading costs down.

Without the roads being in good repair and cleared of snow in winter, the town's residents would find travel difficult, not only by car but also by foot. Marshes cover much of the island, and without roads—with their component causeways, culverts, and bridges—through and over them, life would be mushy indeed. Boats have always been an alternative to walking and driving, but at low tide they can be useless in marsh creeks and even in the back river that separates Arrowsic from its neighboring island, Georgetown. Arrowsic's transportation infrastructure—and the budgetary choices it presents to its road commission and citizens—is a microcosm of that of towns and villages, not to mention cities and states, across America.

The entire state of Maine's biennial budget for the fiscal years 2014 and 2015 included for its department of transportation just a bit over $1 billion, of which $385 million came from federal funds. By contrast, California's single fiscal year 2014–2015 Department of Transportation budget was almost $11 billion, with $4.8 billion attributable to federal funds, and to be spent in half the time. The fiscal year 2015 budget request of the Federal Highway Administration was $85 billion. That's a lot of money, and it's not always spent wisely.

One way of spending money wisely is to look at what it buys over the long term. An asphalt road surface has the advantages that it can be more quickly laid down, it can be opened to traffic use virtually immediately thereafter, it provides a joint-free and thus a smoother and quieter ride, and it is almost always less expensive than the alternative material. A concrete road has its own advantages, however, since concrete can last for many more decades than asphalt, largely because, if properly designed and placed, its surface is more durable and does not develop potholes so easily. The choice between the two road-surfacing materials often rests on the outcome of a cost-benefit analysis done by whoever is in charge of road building and maintenance. Of course, there are many nontechnical factors that can enter into the resulting decision-making process, including preconceived ideas, irrational preferences, politics, business interests, conflicts of interest, and limited budgets, not to mention graft and corruption. Even where local decision making is done in a relatively honest climate—meaning that the pros and cons are weighed openly, taking into account the best interests of the citizens who elect the decision makers and whose taxes pay their salaries and the bills—the choice between asphalt and concrete may not be obvious.

Given a limited budget for roadwork, is it better to choose which few among all the deteriorating roads in a jurisdiction should be paved with concrete that might last for several decades, or to pave many more roads with asphalt, recognizing that they may have to be resurfaced in fifteen or twenty years? It is when faced with such tough choices that elected officials often resort to hiring a consultant who is paid to study the problem and make recommendations. The consultant is, of course, subject to the same kinds of prejudices and temptations that any other decision maker is. So choosing a consultant can be, in itself, a difficult decision fraught with pitfalls.

A good deal of deliberative information about making sound infrastructure choices exists on the World Wide Web, but like much on the Internet, locating it can be difficult. Some very helpful guidance may be found in memos and reports prepared by town or county engineers who had been tasked with laying out the argument for one approach to road surfacing versus another. These documents, written for nonspecialists sitting on a planning board or town council, avoid or define jargon and generally make a complicated matter understandable in relatively simple terms.

Due to a usually lower paving cost, the predominance of American roads are asphalt covered. But because asphalt generally develops potholes more quickly than concrete deteriorates, in the long term the maintenance cost of an asphalt road can be expected to be greater than that of a concrete one. Still, the choice at hand is not always straightforward. This is illustrated along a stretch of I-85 in northern North Carolina and southern Virginia that exposes drivers and passengers to a variety of asphalt and concrete pavements, suggesting that different conclusions were drawn at different times, perhaps by different groups of people advised by different consultants or engineers. Portions of the highway, especially those newly paved in either concrete or asphalt, are in wonderful condition; others are in such poor shape that drivers must wish they were on an alternate route. The last time I drove over it, one section of I-85 in Virginia had so many patches—asphalt on concrete, concrete on asphalt, and patches overlapping patches—that it was impossible to avoid the bumpy transitions. Even switching from the truck to the passing lane provided no relief. The ride felt as bad as the road looked.

In dealing with tight budgets and reelection campaigns, getting something cheaper up front appears to be the more attractive deal. The "first price" of an asphalt road was noticeably lower than that of a concrete one when the price of oil, from which asphalt is of course ultimately derived, dropped significantly in 2014. However, when the Nevada Department of Transportation took long-term costs into account, it opted for concrete in awarding an $83 million paving contract. Although this was $3 million more than the low bid for asphalt, the department applied a "life-cycle equivalency factor," which allowed it to consider the fact that over a thirty-five-year period the cost of maintaining the asphalt would have been three times the initial paving expense. When looked at over the long term, there was little question which was the better buy.

But even after a big infrastructure decision is made, who decides how a brand-new road or bridge will be finally finished in terms of choosing such details as the style and quality of guard rails and lighting fixtures? Such decisions can be assigned to junior staff members who have no experience, expertise, or sense of excellence, and the result can be disappointing, to say the least. In the case of one bridge whose design history I know, the final decision about lighting fixtures was unfortunately put to a vote of the local citizenry, who chose a style that is wholly consistent with the old river town but strikingly incompatible with that of the modern bridge itself. Giving people the ability to participate in the design, albeit only of the appurtenances, of a major piece of infrastructure may be politically savvy but is professionally irresponsible. A well-designed bridge, like a well-designed house, should be consistent in its style. Such consistency cannot be achieved if such finishing details as lighting fixtures are selected in isolation from the design of the bridge proper.

Roads can suffer the same incongruity of design elements if careful holistic thought is not given to how they are lighted, how they are marked, how they are signed, and how they are bridged. Not every road can be a successful museum of overpass bridges like the Merritt Parkway in Connecticut, but introducing variety into such dominant elements of the parkscape as its cross bridges makes it less likely that drivers will be lulled into a semiconscious state of déjà vu as they drive consecutively under such similar-looking crossings that it can appear that no progress is being made in the journey. But roadside elements like light poles are virtually invisible to the typical driver, and their repetitiveness will likely go unnoticed. Still, their design should be compatible with the land-scape, both natural and constructed. Likewise, decisions about how to resurface streets and highways should be thoughtful and consistent technically, aesthetically, and financially.

16

Trodden Black

CORRUPTION, GRAFT, WASTE, FRAUD, ABUSE

A BRAND-NEW ROAD MAY look perfectly constructed, but the illusion of quality may be only surface-deep. Was the foundation properly prepared? Will the road's profile shed rainwater correctly? If it is a concrete road, was the appropriate amount of reinforcing steel used, was the concrete mix correctly proportioned, and were conditions under which it was placed such that the concrete was allowed to cure properly? If an asphalt road, was the old surface properly prepared, was the asphalt used of the required specifications, was it applied in the correct manner to the correct thickness? Was it done on a warm enough day? There is much behind and beneath a roadway surface that is not apparent to casual inspection. To ensure that the road is constructed as designed and specified, oversight throughout the process must be honest and vigilant. Unfortunately, sometimes the entire system can seem to be corrupt.

Stories of fraud and abuse within and outside government are legion, and highway contracts provide a good share of them. A recent example comes from North Carolina, where in 2014 an executive of a road paving company pleaded guilty to charges of conspiracy to defraud the U.S. Department of Transportation and also conspiracy to launder money. The scheme involved misrepresenting the role of a minority-owned contractor in road projects involving thirty-seven federal construction contracts totaling more than $87 million over a decade. None of the contracts would have gone to the paving company in the first place had it not stated that a "disadvantaged business enterprise" would get a certain percentage of the work. In particular, the involvement of a minority-owned trucking company was fraudulently exaggerated in bidding documents and its role later misrepresented by affixing magnetic signs displaying the trucking company's logo over that of the paving company's on its own trucks. In addition, the trucking company's letterhead was used to document contracts and records that were totally false.

The paving company did pay the trucking company about $375,000 for work actually done, but the federal requirement for its involvement was ten times that amount. The difference was placed by the paving company in a phony bank account that bore the trucking company's name, and most of that money got fed back to the paving company. The trucking company's owner was given kickbacks for his cooperation. According to the Department of Transportation's Office of the Inspector General, fraud involving disadvantaged business enterprises had risen in 2013 to almost 30 percent of the office's caseload. The example just cited was only one of about eighty cases involving paving companies that resulted from a sweeping federal investigation that was called "the largest anti-trust probe in U.S. history."

What "the largest anti-trust probe in U.S. history" means in the context of corruption is anyone's guess, for it is hard to estimate how pervasive illegal activities actually are. After all, except for what we might know anecdotally and, heaven forbid, from personal experience, there might be an awful lot of illegal stuff going on that is never detected or perhaps even suspected—or activity that is suspected, detected, but rejected as not being worth getting involved in.

Not long after the North Carolina example, I became aware of a case involving the rebuilding of the World Trade Center in New York City. The owner of a Canadian company that was responsible for almost $1 billion worth of contracts for the project was arrested and charged with having engaged in fraud by claiming that minority- and women-owned businesses were doing work that they were not: Over the course of three or four years, false payroll records were submitted to the building's owner, the Port Authority of New York and New Jersey. There can be a lot more behind a striking architectural façade than a steel structural frame, and the phenomenon is not new.

One revealing example of corruption took place in Washington, D.C., in the 1870s. Alexander Robey Shepherd, a native Washingtonian, worked his way up from being a plumber's helper at age thirteen to governor of the district, which made him also head of its Board of Public Works. His influence as a mover and shaker earned him the nickname "Boss" Shepherd. As a means of countering the financial depression of the time, and of improving the dismal condition of Washington's infrastructure, Shepherd's administration floated a $5 million bond issue to install a sewer system and lay wood-block pavements, the latter known to be quick and cheap to implement. However, politically chosen private entrepreneurs were contracted to do the work; they were without the proper knowledge and skills and laid sewer lines that ran uphill and pavements that soon rotted away. The waste of resources led Congress to seize control of district governance and place it in the hands of a board of commissioners. The failed sewer system and road pavement were rectified by the Army Corps of Engineers. Fixing the former involved a lot of digging up and relaying pipe; fixing the latter involved using asphalt to resurface the streets. In the wake of Boss Shepherd's reign, paving projects generally were viewed with skepticism. Boss Tweed, Shepherd's near contemporary who ruled New York's Tammany Hall, knew better than to allow inferior paving jobs to threaten his dominion. Tweed did not skimp on the quality of roadwork because he knew that failing pavements could give his political opponents a great advantage in the next election.

Norman Bel Geddes, in a chapter entitled "Eliminate Graft and Double Highway Construction" in his 1940 book *Magic Motorways*, made a self-described "strong statement": "Twice as much money is spent for roads today as is justified by results." He defended the claim as resting on two facts, namely, "inefficiency and graft," and admitted that it was "often difficult, if not impossible, to distinguish one from the other," since "graft is often passed off as inefficiency." Among the more subtle forms of inefficiency he named were waste and improper construction, resulting in "roads that cost too much to build, wear out too quickly, require constant repair and unnecessary maintenance costs." Though Bel Geddes' use of the multiplier "twice" to describe the amount spent might be argued, much the same might be claimed today. It is highway robbery, and it is nothing new.

In the early twentieth century, long before Thomas Harris MacDonald became chief of the Bureau of Public Roads, he was just starting out as a young engineer working for his mentor Anson Marston, who had only recently been appointed dean of the new division of engineering at Iowa State College, which Iowa legislators had designated a state highway commission. That was in 1904, when the entire state contained fewer than a thousand automobiles, so unremarkably MacDonald rode on horseback to visit construction sites. What he found was an abundance of shoddy and wasteful work, with local communities often getting "about a dime's worth of road for every dollar they spent." When he looked at bridges, he found they were more than likely wooden and not well built at that. The state had been carved up among contractors so that each of them could capture whatever bridge work there was to be done in a particular area. This kind of arrangement caused the taxpayer to pay twice—once in overpriced contracts and then again in construction that did not last.

Deceit can take the form of highway funds being diverted to other purposes, leaving a diminished balance for the intended purpose. Money from bonds issued for roadbuilding may not be spent as promised, resulting in inferior design and construction. In time, according to the visionary Bel Geddes, writing in 1940, "The citizens who used and paid for the roads began to realize how thoroughly they were cheated when many miles of road crumbled after just a year of use."

Not all malfeasance necessarily leads to shoddy work. Bel Geddes traced the concept of highway robbery back to the days of primitive trade routes, and gave some compelling examples from the nineteenth century. One case involved the infamous Boss Tweed, "New York's master corruptionist," who has been described as the very personification of political wrongdoing. According to Bel Geddes, Tweed came up with a scheme to build along the dramatic New Jersey Palisades a "monumental boulevard" leading to a scenic hotel on Hook Mountain, located just north of Nyack. Beginning in the 1870s, aided by the use of the newly developed dynamite and heavy construction equipment, extensive quarrying of the nearby cliffs yielded crushed stone suitable for use in macadam roads well beyond the projected boulevard. The stone was also later employed as aggregate for concrete used in building construction in New York City. Such activity allowed Tweed-related contractors to profit from use of the stone long before the boulevard project even began. The destruction of the cliffs finally ceased with the incorporation of the quarry sites into protected recreational and parkland.

Overpaying for the maintenance of road or other infrastructure has been a blatant form of the waste of public funds. In 1938, an Ohio highway department engineer testified in front of a state senate committee about padded estimates for "bituminous road material"—the stuff used to make asphalt pavement. Whereas the federal government bought the material for $6.56 per cubic yard, Ohio had been paying $14 for it. In related testimony, it came out that someone at the Democratic Party headquarters had to be paid twenty-five cents for every ton of the material supplied by one company. The engineer who testified about the overestimating was fired by order of the governor.

In the late 1960s in Maryland, engineering firms selected to work on state construction contracts were not chosen through a conventional public bidding process but at the discretion of the governor and members of the State Roads Commission, which made for a system that was rife with corruption. It was the practice at the time that kickbacks were paid according to the following formula: 25 percent of the kickback amount to the state official arranging the deal leading to the contract; 25 percent to the state official bringing the deal to the governor; and 50 percent to the governor himself. Some of the deals were said to be made directly with the governor, who at the time was Spiro T. Agnew. He had begun his political life as a Democrat but won election as Baltimore County executive and governor as a Republican. In the 1968 presidential race he was Richard Nixon's running mate and was elected the thirty-ninth vice president of the United States. In that office, he continued to receive kickbacks on Maryland construction contracts. In spite of such behavior, and amid the confusion of the Watergate scandal, Nixon and Agnew were nominated to run for a second term and were reelected in 1972. The kickback scandal involving Agnew became public in the summer of 1973, and as part of a plea bargain agreement he resigned the vice presidency in the fall. (Had he not done so, of course, he would have become president when Nixon resigned the following summer.) So corruption has a way of reaching the highest levels of government. This is not to say it is missing from the lowest.

I recently visited a retired architect who had designed and built his own unique retirement home on the coast. As he showed me around the house, I asked if a detail of the doorway to the basement conformed with the local building code. He laughed and said in effect that there were no building codes because no one enforced them in that location. In elaborating on his experience with building codes, he used the word "wise guys," which reminded me of its use in the title of the book, *Good Guys, Wiseguys, and Putting Up Buildings: A Life in Construction*, by the engineer and writer Samuel C. Florman, who practiced in the New York City metropolitan area beginning in the 1950s.

In a chapter titled "Corruption," Florman relates a story of when he was just starting out in the construction business. One day his boss told him to go to the Department of Buildings to obtain a certificate of occupancy for a project that had just been completed in Queens. Final inspections had been done, and the paperwork was ready to be distributed to the appropriate individuals. He was instructed to stop at various desks, sign papers, and pick up documents. Before leaving on his assignment, he was given some five-dollar bills, one or two of which he was to drop into a desk drawer when the person at the desk opened it. The young and idealistic Florman questioned his boss about bribing city employees but was told that the money was just a tip. When Florman replied that even giving tips was illegal, he was told that everybody did it in the real world of New York construction. In a 2008 *New York Times* article on the long history of graft and corruption in the city where about $12 billion in private construction takes place each year, the buildings department was termed "echt New York," with the only thing distinguishing it from other city departments being "its ostensible invulnerability to reform."

But back to roads, where reform does seem to happen now and then, at least for a while. The construction of the Merritt Parkway in Connecticut, which opened in the late 1930s, was riddled with corruption that was detailed in a report made to Governor Wilbur L. Cross. According to the report, as stated breathlessly by Bel Geddes, "Highway contracts were rigged. Miles of roadside graded by one highway unit were torn up by another. Cracked concrete bridges were approved. Road materials were inadequately tested, and credibly enough, the work was poorly inspected. Land assessed at about $14,000 was sold to the state by one of its legislators and members of his family for $100,000! In Greenwich, the state paid over $1,000,000 for land assessed at less than one-tenth that sum." At least some of the miscreants responsible for such flagrant violations of good citizenship, not to mention of the law, got caught. Following a grand jury investigation, a state highway commissioner resigned at the request of the governor, and land acquisition proceeded in a more equitable fashion.

Today, the Merritt Parkway is a continuation of New York's Hutchinson River Parkway (named for the Puritan spiritual advisor Anne Hutchinson) and runs for almost forty miles from the state line at Greenwich to the Housatonic River at Stratford, where the road becomes known as the Wilbur Cross Parkway. This trio of parkways running tandem between the New York City borough of the Bronx and Meriden, Connecticut (located about fifteen miles south of Hartford), provides the long-distance automobile driver a welcome respite from the interstates and their trucks. The Merritt, which is named for U.S. congressman Schuyler Merritt, who championed its enabling legislation, when new was hailed as a model of parkway landscaping and architectural bridge design. In 1991, it was added to the National Register of Historic Places. However, barely two decades later it was named one of America's Most Endangered Historic Places by the National Trust for Historic Preservation. The ongoing erosion of the Merritt's most distinguishing features led to the establishment in 2002 of the Merritt Parkway Conservancy, which is dedicated to the protection and preservation of

No two overpasses on the Merritt Parkway in Connecticut look the same, making the drive along the parkway through Fairfield County a far-from-monotonous experience. This overpass is the Redding Road Bridge; it reflects the imagination and creativity of the planners and designers of this stretch of infrastructure.

the parkway's original character and integrity. Unfortunately, tight budgets, lack of appreciation for the landmark's historic significance, and apparent preference for function over form by the Connecticut transportation department have made the efforts of the conservancy ones of constant battle.

Still, money intended for but not used for highway improvements ultimately comes out of the pocket of the citizen taxpayer. Those who cheat, lie, and steal are modern versions of the highwaymen who lay in wait beside the trade routes to hold up travelers and traders. In the twentieth century, being waylaid took many forms, including diverting funds to other than their intended purpose. Bond issues have long been used to raise money for roads, bridges, and other forms of infrastructure, and the bonds have ultimately been backed by the citizens who should benefit from the improvements. However, in example after example, the proceeds of a bond sale were not fully applied to the road construction for which the bonds were sold in the first place. Instead, the funds were diverted to other projects entirely, the money was distributed in the form of graft, or the amount spent simply did not match the quality of product received.

In an instance in 1930s Arkansas relating to $163 million raised through the sale of bonds, the governor expressed the opinion that "a fair appraisal of our roads will show a 50 per cent value of the bond issues." The influx of out-of-state money had been greeted with delight, but in the elated atmosphere accompanying the windfall there was insufficient thought given to how the bonanza would be administered and accounted for, or how work done with the funds would be overseen and monitored. In time, however, the citizens who had paid for and used the roads realized how much they were cheated when mile after mile of pavement crumbled after about a year in service. And the miles were not bought on the cheap, for an audit commission found after the fact that on a single $10 million job there had been overcharges of more than $4 million. That's a 67 percent premium that evidently produced nothing of public value. And that was just one contract in one state. According to Chester H. Gray, director of the National Highway Users Conference at the time and a longtime proponent of a nationwide system of highways and the proper maintenance of existing roads, the practice was not limited to Arkansas and "states which are now guilty of diversion of highway funds see their roads deteriorating."

There would be echoes of the 1930s in the early years of construction of the interstate highways. In the late 1950s, accusations of corruption and waste, including in the form of fraud and land speculation, led to investigations by the House Special Subcommittee on the Federal-Aid Highway Program, chaired by Minnesota Representative John Blatnik, who once maintained that corruption "permeates the highway program and stigmatizes the whole road-building industry." To illustrate the pervasiveness, Walter May, the committee's chief counsel, proposed throwing a dart at a map of the United States. "Wherever it sticks," he asserted, "we can find something wrong with the new highways."

At the beginning of interstate construction in 1956, the Bureau of Public Roads, which was overseeing the program, had projected the final cost to be $27.5 billion; after the first five years of construction the revised estimated cost was $41 billion. Some of the 50 percent increase was attributed to inflation, but an estimated 10 to 20 percent of it was blamed on graft amounting to at least $100,000 per mile of highway built. The committee did verify examples of offenses, but Blatnik later held that the areas in which faults were found constituted "only a small fraction of the total of this great program." Nevertheless, the Bureau of Public Roads did establish an Office of Audit and Investigations.

Fortunately, the horror stories of corruption, graft, waste, fraud, and abuse are remarkable because they are not the norm, and today's highways, even though receiving poor report card grades overall, are clearly not all deteriorating. Unfortunately, in my experience, enough of them may be in poor enough condition to make drivers and voters wonder if we might well be back in 1930s Arkansas or in the earliest years of interstate highway construction. It may indeed be a small fraction of road contracts that are not what they should be, but all it takes is a couple of bad anecdotes to lead to a barrelful of generalizations.

17

For Another Day

FUEL TAXES, TRUST FUNDS, AND POLITICS

PRIOR TO LEGISLATION KNOWN as the Federal Aid Road Act of 1916, funding by Washington of highways other than post roads was extremely limited. Indeed, it has been argued that the federal government had no constitutional authority to build roads itself, and so the responsibility had fallen upon each individual state—and its counties, cities, and towns—to plan, design, and oversee the construction of roads and highways (including bridges) within its boundaries. Where a bridge crossed a state line located in a river or other navigable waterway, a federal concession had to be granted, but except for specifications about width of channels and height of clearances, the central government stayed out of the bridge design and building process. However, as is the case today, states and businesses did not stay out of lobbying efforts associated with the design, drafting, passing, and enacting of laws: the 1916 law was based on model legislation developed by the American Association of State Highway Officials, and at the signing ceremony President Woodrow Wilson was joined by members of AASHO, the American Automobile Association, and farm organizations.

Wilson was an advocate of good public roads for rural and urban areas alike. The federal aid act provided for fifty-fifty matching funds, and to secure them for a proposed project a state had to submit surveys, plans, specifications, and estimates to the secretary of agriculture demonstrating how important the cost-sharing scheme was for developing roads by which farmers could transport their produce to cities.

While the federal government may not have had much to say officially about the details of roads, the secretary's approval or rejection of a request for federal funds sent a signal to the states as to what kinds of road projects were preferred by Washington.

World War I gave citizens and legislators alike other things than roads to think about. But in the wake of the war America's roads were brought back into focus by the cross-country caravan in which Lieutenant Colonel Eisenhower and others noted the sad state of the country's transportation infrastructure. The 1920s seemed to distract almost everyone from everything, and then came the Great Depression; but, throughout it all, highway planning continued. The Federal Highway Act of 1921 was described by Earl Swift in *The Big Roads* as "the single most important piece of legislation in the creation of a national network," the interstate legislation thirty-five years in the future notwithstanding. The 1921 act mandated a true national system of highways. Although the individual states were the designated planners, the federal government, through its power of the aid purse, enforced standards of design, construction, and maintenance. However, the federal funds were not always fully available.

To deal with the Depression, President Herbert Hoover sought to balance the federal budget by simultaneously reducing federal spending and increasing revenue. The latter was aided by the Revenue Act of 1932, which placed new taxes on everything from income to toiletries. Among the levies was a one-cent-per-gallon tax on gasoline, which was expected to bring in about $150 million a year. But this revenue was to be used for reducing the deficit, not for improving highways. The tax was evidently just something easily calculated and the revenue easily delivered to the government. The Depression kept the federal financial support of highway construction on the back burner.

However, World War II brought new problems. President Franklin Roosevelt recognized in the midst of it that when the war was over, millions of returning soldiers would have to be absorbed back into the economy. In addition, industries would have to convert from war production back to peacetime work. Among other things, a committee was appointed to revisit the 1939 report *Toll Roads and Free Roads*, which led to an extension and refinement of its ideas in a new report titled *Interregional Highways,* which was completed early in 1943. FDR submitted it to Congress the following year in the hope that it would "facilitate the acquisition of land, the drawing of detailed project plans, and other preliminary work which must precede actual road construction."

Congress was favorably inclined to do so but used the term "interstate" rather than "interregional" in the Federal-Aid Highway Act of 1944. The legislation declared, among other things, that "there shall be designated within the continental United States a National System of Interstate Highways." The system was to be "so located as to connect by routes, as direct as practicable, the principal metropolitan areas, cities, and industrial centers, to serve the national defense, and to connect at suitable border points with routes of continental importance in the Dominion of Canada and the Republic of Mexico." However, the bill made no specific provision as to how an interstate system would be financed.

In the wake of World War II and the Korean War, the matter of funding highways had to be confronted. From a period of exponential growth into the 1920s, automobile registrations had peaked at 23 million in 1930 and actually dropped in the early years of the Depression. Truck registrations likewise began to stall, peaking at about 3.6 million. Significant growth resumed after World War II, and by the mid-1950s there were registered about 53 million automobiles and 12 million trucks ready to use highways built for a much smaller volume of traffic. Advocates for improved highways and their financial support continued to object to revenue generated by taxes and fees on motor vehicle drivers and owners being thrown into a general government fund. The gasoline tax, which in the early 1950s stood at two cents per gallon, was opposed by many governors, who much preferred individual state-controlled programs to a single federal-aid highway program. In contrast, however, state highway officials feared that doing away with the federal tax on fuel would not necessarily mean that the revenue from it would devolve to the states, because there would be political pressures not to replace the federal with a state fuel tax. These officials came to favor a system whereby the federal tax revenue was tied directly to federal highway aid to the states, and AASHO lobbied for this model.

However, under the new model, the federal government provided in the form of aid to the states only a portion of what it collected in gas taxes. During the 1952 presidential campaign, candidate General Eisenhower argued that "a network of modern roads is as necessary to defense as it is to our national economy and personal safety." At the signing ceremony for the Federal Aid Highway Act of 1954, President Eisenhower noted the "rough linkage of gas tax revenue and highway authorization" that the legislation represented. He also characterized it as a step forward in meeting the nation's considerable highway requirements that were becoming increasingly severe.

The needs not fully addressed in the 1954 legislation were not new to legislators. In the Federal-Aid Highway Act of 1944 a forty-thousand-mile National System of Interstate Highways had been authorized, but most states had chosen not to use federal aid funds to implement a federal preference. They had instead used the money for their own priorities. Understanding this, in 1955 Eisenhower put forth to Congress his plan to build in ten years an interstate highway system that would be financed by $30 billion in bonds backed by federal gas tax revenue. Congress did not like the debt aspect of the plan but did come to like the idea of a user tax that would allow for a pay-as-you-go method of financing the interstate highway system. The Federal-Aid Highway Act of 1956 authorized a National System of Interstate and Defense Highways, with the details to be worked out in committee. The resulting highway revenue act estimated the cost of the system at $35 billion over fifteen years, with all revenue from highway user taxes (3 cents per gallon on gasoline, plus increased taxes on other fuels, truck tires and tubes, etc.) dedicated to highways through a Highway Trust Fund. The federal government controlled the fund, dispensing matching monies from it to provide for state and local road and bridge projects it deemed worthy of support. Not surprisingly there were many other purposes to which legislators and lobbyists wished also to put it.

By the early 1960s, the federal gas tax was increased to four cents per gallon; in 1983 it jumped to nine cents, with one cent of that going to a

new Mass Transit Account. In 1990, as part of legislation to reduce the
federal budget deficit, the gas tax was increased by five cents, only half
of which went into the Highway Trust Fund. A second Omnibus Budget
Reconciliation Act added 4.3 cents to the tax, all of which went
to deficit reduction until 1997, when it was redirected to the Highway
Trust Fund. These additions, plus a 0.1-cent tax added in 1986 to
fund the remediation of leaking underground storage tanks, brought
the total federal gasoline tax to 18.4 cents, where it remained in the
mid-2010s.

For decades the highway trust fund had been stable and its balance
growing, but when in the early twenty-first century people began to
drive less, in part because of increased fuel prices and in part because of
the poor economy, there was of course less gas tax revenue to feed the
fund. The causes of the fiscal problem were generally agreed upon. The
federal gasoline tax had been fixed, with no provision for inflation, at
18.4 cents per gallon (24.4 cents for diesel fuel) since the Omnibus
Budget Reconciliation Act of 1993. Since this tax was the single largest
contributor to the trust fund, its balance was not keeping up with inflation, let alone with increasing infrastructure needs. Revenue was also
declining as motor vehicles became more efficient, using less fuel,
and the growth in the use of hybrid and all-electric vehicles further
reduced fuel consumption and hence fuel tax revenue. These unantici-
pated consequences of economy and technology threatened the health
of the Highway Trust Fund.

In a 2014 article on the American transportation infrastructure, the
Los Angeles Times reminded readers once again that of the nation's
approximately 600,000 bridges, almost 150,000 were considered deficient or obsolete and an infrastructure that was "once an engine of
mobility and productivity . . . has fallen into such disrepair that it's
become an economic albatross." The nation's roads and bridges were
called "shoddy," a claim supported by the fact that in California about
one-third of public roads were not in good shape and that in the most
recent fiscal year New Jersey received almost 10,000 reports of potholes,
more than twice the number for the previous year. According to a report
from the Pew Charitable Trusts—the first in its *Fiscal Federalism in*

Action series—between 2007 and 2011 the average annual spending on highway and transit systems nationwide was $207 billion, of which the states provided $82 billion (or 40 percent), localities $74 billion (35 percent), and the federal government merely $51 billion (25 percent). However, the relative percentages varied greatly from state to state, with the federal government contributing 55 percent of the total in Montana but only 15 percent of it in New York. Regardless of the breakdowns, overall spending was down and sources of revenue were not keeping up with needs.

It had been clear well before the summer of 2014 that something had to be done about spending and funding for roads and bridges. During that spring the testimony before the U.S. Senate Committee on Finance of a representative of the Congressional Budget Office included presentations on "The Status of the Highway Trust Fund and Options for Financing Highway Spending" and "The Highway Trust Fund and Paying for Highways." At the time, Congress was facing what essentially should have been the three perennial questions about the federal government's involvement in the nation's transportation infrastructure: How much should be spent on highways? How should those funds be raised? How should the use of those funds be directed?

At the time, the United States was spending on highways alone a total of about $160 billion a year, with the federal government nominally contributing about $40 billion of that. This 25 percent share was raised mainly through the gasoline and diesel fuel taxes that drivers pay at the pump. However, for years Congress had been spending more on highways and surface transit systems than it was taking in in fuel taxes and fees. In a typical recent year, spending for transportation infrastructure had been about $50 billion, but dedicated revenue only about $34 billion. Since 2008 the general fund of the U.S. Treasury had been tapped for the difference, amounting to a total of $54 billion over that period. The expectation for the fiscal year 2015 was that $18 billion would have to be transferred into the trust fund. Projections were that the shortfall would grow to $120 billion over the next decade.

The then current transportation funding bill was scheduled to expire in September 2014, at the end of the federal government's fiscal year.

Everything came to a head that summer, as the date certain for bank-ruptcy of the Highway Trust Fund was clearly August. Unless some-thing was done before then to change the trajectory of spending or funding, a dangerous fiscal intersection would be reached and the former would overtake the latter. Federal funding for highway and bridge work would have to be decreased, leading to the cutting back or outright cancellation of construction projects, with the loss by layoffs of as many as seven hundred thousand construction jobs associated with those projects, adversely impacting the nation's overall economy, which was not strong to begin with. The Department of Transportation let it be known that in the absence of saving legislation, it would have to begin cutting back by as much as 30 percent on payments to the states.

Among the obvious solutions to the shortfall problem was to raise the gas tax immediately. But November midterm elections were approaching, and every one of the representatives and one-third of the senators were up for reelection. None of them wanted to vote for a rise in the tax on fuel—or to shut down temporarily or cancel completely construction projects—in an election year, and so they looked for other sources of revenue for a quick fix. If the trust fund could be put in the black until after the elections, a long-term solution to the problem might be worked out then. Yet, as they had been for almost four years on virtually every issue that had come before Congress, the Republican-led House of Representatives and the Democrat-led Senate were not in agreement on how to proceed to a resolution. The highway trust fund matter approached crisis proportions as August neared.

In mid-July, the House had passed a bill designed to add $10.8 billion to the trust fund, which would keep it solvent until the following May, the start of a new construction season. Among the sources of money was a transfer of $1 billion from a distinct fund set aside for cleaning up polluted sites, like abandoned service stations, contami-nated by leaking underground storage tanks. Since most such work had already been completed, this fund had been running a $200-million-per-year surplus. A second contributor was to be an extension on customs fees paid on vehicles and other goods brought into the country, amounting to $3.5 billion. Finally, a so-called gimmick termed "pension smoothing" was proposed in which corporations having defined benefit retirement plans for their employees could, by assuming higher interest rates in the future, lower their required tax-deductible contributions to the pension funds, thereby raising their federal income taxes and so increasing revenue to the government. Presumably, especially if the interest-rate assumption was overly optimistic, a company would add to an underfunded pension account years later, in which case it might pay lower taxes then. Since the federal government estimates the cost of legislation only over the coming ten years, the long-range impact of that lower revenue to the U.S. Treasury was not taken into account. It appears to be the nature of congressional logic to consider the world illogical. In any case, pension smoothing was estimated to raise $6.4 billion. The three prongs of the congressional action, allowing for rounding, added up to the $10.8 billion shortfall in the highway trust fund.

When the bill was sent over to the Senate, it not surprisingly met with opposition and derision, as senators termed it a gimmick. As reported in the *Congressional Record* for July 28, Senator Sheldon Whitehouse of Rhode Island called it "woefully inadequate" and "a pathetic measure" that was "a joke that does nothing on long-term investments in our infrastructure, nothing in a sustainable way to pay for them." Senator Ron Wyden of Oregon expressed his displeasure with the proceedings by declaring, "The reality is that it is just not possible to have a big league quality of life with little league infrastructure." On his Web site, Senator Patrick Leahy of Vermont characterized the short-term extension bill "another artificial, made-in-Congress crisis" representing the "my way or the highway" governing approach. What the Senate ended up agreeing to was an amendment raising the $10.8 billion by somewhat different means, including $3.9 billion from customs fees and transfers from the storage tank fund and $2.7 billion from a pension-smoothing provision that extended over only three years. The balance ($4.2 billion) was to come from tightening tax rules on reporting mortgage interest, extending the statute of limitations on capital gains transactions, and withholding Medicare payments to delinquent taxpayers. Such is the way the government works. As with so much federal legislation, this and the House measure were full of optimistic projections and wishful thinking.

But the most striking difference between the House and Senate bills was the date to which they would extend contributions to the trust fund. Whereas the House provided for help through May 2015, the Senate would prop up the fund only through mid-December 2014. Its idea was to force a lame-duck Congress to vote on more long-reaching legislation, making the trust fund solvent for six years at least. The argument for this was that the states and private contractors could better plan for the future; the argument against was that lame-duck legislators would not be accountable to the voting public. House leadership stated clearly that it was going to stick with its preference for a May extension date and promised to send its original bill to that effect back to the Senate.

The situation with the two divergent bills on the table and the deadline looming recalled the game of chicken like the one dramatized in the film *Rebel Without a Cause*. The *Washington Post* characterized the impending congressional disaster as the "highway cliff," alluding to the "fiscal cliff" that Congress had wrestled so close to in previous years. This was very much the metaphorical situation on the last day of July, with Congress due to recess on the first of August for a five-week vacation, the same day the Department of Transportation was scheduled to begin cutting payments to the states from the Highway Trust Fund. As the debate came down to the wire, the Congressional Budget Office, which "scores" or audits draft legislation to ensure that the numbers are realistic and that they reconcile, found what was described as a "critical error" amounting to an approximately $2.4 billion shortfall in claimed revenue. This embarrassment was effectively a showstopper for the Senate-amended bill, which the House rejected. As promised, it passed the original version back to the other chamber. On the deadline night, the Senate—perhaps because the senators already had their bags packed and were ready to hit the road—relented and voted in support of the almost $11 billion House bill. The Highway Trust Fund was thus replenished into the next spring, and Congress could claim that more than 660,000 jobs and about 6,000 state surface transportation projects would continue without interruption.

In a statement issued the evening that the bill passed, the secretary of transportation, Anthony Foxx, welcomed the good news that Congress had rescued the Highway Trust Fund from near bankruptcy but lamented the bad news that there was "still no long-term certainty, and this latest band-aid expires right as the next construction season begins." He announced the commencement of "a nationwide virtual town hall on transportation" among whose purposes was to engage American citizens to push Congress to action on multiyear legislation that would provide certainty to both businesses and governments for planning purposes. Another crisis had been averted in Washington, but as Senator Angus King, Independent of Maine, observed about the Highway and Transportation Funding Act of 2014, it was just another punt and it "showed an inability to face a real problem and deal with it."

Eventually the real problem would indeed have to be faced, and some of the unsuccessful proposals put forth in the hours, days, and weeks before the punt may in due time come into play. Among other proposals that had been offered was one from Senator Mike Lee, Republican from Utah, to reduce gradually over five years the federal gasoline tax from 18.4 cents to 3.7 cents per gallon. The approach of returning responsibility for highway infrastructure to the states has come to be known in Washington and other political circles as "devolution." It would, of course, ultimately take the federal government largely out of the road-funding business. The individual states would assume virtually complete authority for their road networks, including the interstate highways within their borders. The greatly reduced amount of revenue to the federal government would be used to fund block grants to the states, which could levy their own taxes on fuel and otherwise raise additional revenue for roads and bridges. According to Senator Lee, "Under this new system, Americans would no longer have to send significant gas-tax revenue to Washington, where politicians, bureaucrats, and lobbyists take their cut before sending it back with strings attached." Furthermore, "states and cities could plan, finance, and build smarter and more affordable projects."

That may be true, but some of those entities would have first to climb out of a very deep fiscal gorge. New Jersey, for example, which was long known for having one of the country's lowest gasoline taxes, had for decades financed road and bridge upgrades and maintenance by borrowing money. This created considerable debt service, exceeding $1 billion in 2014. To keep up with this took all of the revenue from the gas tax. The state's transportation trust fund was expected to run out of money before the end of fiscal year 2016, when New Jersey would surely have to make some hard decisions, most likely without the help of the federal government. Texas had $23 billion in debt attributed to road projects and was expected to spend $31 billion over twenty years to pay it off. Funding for California's transportation needs beyond basic road repair and maintenance was $300 billion short over the coming decade.

How the states would actually solve their fiscal problems remained to be seen, but it was very likely that there would be a wide variety of approaches, for the states have long been divergent in their ways and means about roads and related matters. In 1919, Oregon became the first state to impose a gasoline tax, with Colorado and New Mexico following soon thereafter. In July 2015, state taxes and fees on gasoline ranged from a high of 51.60 cents per gallon in Pennsylvania (with New York next highest at 45.99 cents) to a low of 11.30 cents in Alaska. New Jersey's 14.50 cents ranked next to the bottom, and not only in this regard was New Jersey odd, for at the time it—along with Oregon—retained its ban on self-service gasoline pumps. The ban, dating from 1949 legislation, meant that gas could be pumped only by licensed attendants, a requirement that accounted for some fourteen thousand jobs in the state. Other stated advantages of the practice included convenience for the disabled and elderly as well as fire and environmental safety. Replacing a gas tax with some kind of alternative tax will likely be as difficult as changing state-based gasoline station practices that have prevailed for more than sixty-five years.

In the closing months of 2014, due in part to increased supplies, average gasoline prices dropped to the lowest levels in five years. Some legislators saw the relatively low prices as presenting an opportune time to raise taxes without their even being noticed. Some U.S. congressmen

TOTAL STATE TAXES AND FEES ON GASOLINE, JULY 2015 (CENTS PER GALLON)

Alabama 20.87	Illinois 35.99	Montana 27.75	Rhode Island 34.00
Alaska 11.30	Indiana 32.89	Nebraska 27.00	South Carolina 16.75
Arizona 19.00	Iowa 32.00	Nevada 33.85	South Dakota 30.00
Arkansas 21.80	Kansas 24.03	New Hampshire 23.83	Tennessee 21.40
California 42.35	Kentucky 26.00	New Jersey 14.50	Texas 20.00
Colorado 22.00	Louisiana 20.01	New Mexico 18.88	Utah 24.51
Connecticut 40.86	Maine 30.01	New York 45.99	Vermont 30.81
Delaware 23.00	Maryland 32.10	North Carolina 36.25	Virginia 22.33
District of Columbia 23.50	Massachusetts 26.54	North Dakota 23.00	Washington 37.50
Florida 36.42	Michigan 35.99	Ohio 28.00	West Virginia 34.60
Georgia 32.62	Minnesota 28.60	Oklahoma 17.00	Wisconsin 32.90
Hawaii 44.76	Mississippi 18.78	Oregon 31.07	Wyoming 24.00
Idaho 32.00	Missouri 17.30	Pennsylvania 51.60	

Source: American Petroleum Institute, http://www.api.org/oil-and-natural-gas-overview/industry-economics/fuel-taxes.

advocated raising federal taxes on fuel, and some states actually did raise their portion of fuel taxes effective New Year's Day 2015. These were not long-term solutions, however.

As Highway Trust Fund legislation neared expiration at the end of May 2015, Congress once again wrestled with extending it. Early in the week before the Memorial Day weekend the House passed a two-month extension, and on early Saturday morning, as they were heading off for a holiday recess, the Senate did the same. This was small consolation for state departments of transportation wrestling with whether to cancel highway contracts, and it only ensured that Congress would once again have the opportunity to kick the can farther down the road at the end of July, when members would be leaving town for their August vacation. There was no reason to think that they would not eventually grant yet another short-term extension of the solvency of the Highway Trust Fund, since in the previous six years or so, Congress had voted almost three dozen times for a temporary extension rather than for a long-term infrastructure bill that would enable states to plan, begin, and complete construction projects with some certainty that the money would be there to pay for them. The new deadline became October 29, 2015.

One possible alternative to the per-gallon tax is to replace it with a sales tax, thereby tying the tax to the price of fuel—and therefore to inflation—rather than to the volume of fuel purchased. And still another alternative to the traditional federal gas tax that had been under consideration for some time was to do away with a tax based on the amount of fuel consumed and convert to a system based on how many miles a vehicle is driven, which is one way to deal with the uncertainty of future tax revenue because of increasing fuel economy and the rise in the use of electric vehicles. However, the scheme presented the problem of how to determine how many miles a particular vehicle would be driven each year. It could be done with a GPS-based tracking system, of course, but some considered that a potential violation of privacy. An alternative would be to rely on self-reported annual miles driven, but would that lead to any more certainty of tax revenue?

Oregon, the pioneer in taxing gasoline, in 2015 became the first state to establish a major volunteer mileage-based program. Those participating in the program were able to choose among a number of mileage monitoring devices and means of reporting, not all of which depended upon a tracking system that was satellite-based. Participants agreed to pay 1.5 cents per mile of road usage and would also receive monthly rebates for fuel taxes paid. It was all but certain that other creative revenue-enhancing solutions would be found by the many other states that were looking into charging for per-mile road use. It is equally likely that creative evasive measures would be dreamed up by taxpayers.

18

How Way Leads on to Way

ECONOMICS, WRONG TURNS, AND POLITICAL CHOICES

IN THE SUMMER OF 2014, the jobs numbers and unemployment figures released monthly out of Washington, D.C., were interpreted by optimists as indicating that the nation was continuing on the road to economic recovery and that its fiscal future looked bright. Those assessments did not square with reality for citizens who held a mortgage for more than their house was worth, who reluctantly had allowed their adult children to move back in with them, and who feared that their grandchildren would have a lower standard of living than they. Nor did the assessments square for the adult children, who continued to be under- or unemployed, who remained mired in student loan and credit card debt, and who felt pessimistic about their prospects for a better life. Everything seemed to constrain them from reaching even a plateau from where they could pause and gain some perspective on what possibilities were available to them in their life and career.

In the midst of this middle- and lower-class malaise, Paul Krugman declared in one of his op-ed essays for the *New York Times* that the nation's economic problems were not hopelessly complicated and mysterious. Rather, he argued, citing the macroeconomist Dean Baker, the explanation for the cause of the difficulties that was coming to be known as the Great Recession was "almost absurdly simple." Krugman stated his clarification as follows: "We had an immense housing bubble, and, when the bubble burst, it left a huge hole in spending. Everything else is footnotes."

So, is the absurdly simple explanation that a gigantic frost heave developed underneath Spending Street? And when it lifted the pavement, had it created conditions conducive to fiscal pothole development? And had a gigantic pothole slowed to a crawl traffic headed toward Recovery Road? Did consumers in their nightmares abandon their cars and become pedestrians, hoofing their way to stores that were forever out of reach and were collapsing under the weight of their own inventory? Was this "Why America gave up on the future," as stated in the large-type teaser accompanying Krugman's column?

After explaining how simple the explanation for the recession was, Krugman turned to what government could do to solve the problems accompanying it. His answer was that "the appropriate policy response was simple, too: Fill that hole in demand." So giant potholes had slowed consumer traffic not only in Spending Street but also in the middle of Demand Drive. Getting around their intersection was going to be rough going indeed.

As Choate and Walter had argued in 1981 in *America in Ruins*, Krugman was calling in 2014 for increased spending on infrastructure as a way to beat the recession. His main observation and recommendation was that "the aftermath of the bursting bubble was (and still is) a very good time to invest in infrastructure." In times of prosperity, he argued, "public spending on roads, bridges and so on competes with the private sector for resources," but since 2008 "our economy has been awash in unemployed workers (especially construction workers) and capital with no place to go (which is why government borrowing costs are at historic lows)." So, "putting those idle resources to work building useful stuff should have been a no-brainer."

Instead of an increase in infrastructure investment to fill the demand hole, however, government responded to the Great Recession with "an unprecedented plunge in infrastructure spending." In spite of highly visible and touted stimulus programs, complete with large signs advertising where the funding for highway projects was coming from, by Krugman's calculation, "adjusted for inflation and population growth, public expenditures on construction" fell by more than 20 percent from early 2008 levels. He characterized the downturn as "an almost surreally

awful wrong turn" that had both short- and long-term negative effects
on the economy and its prospects for the future. And, he warned, it
looked like things were going to get still worse. He was writing this as
Congress was battling over what to do about the Highway Trust Fund
as it approached bankruptcy. Without the temporary fix that Congress
finally enacted at the eleventh hour, Krugman reminded his readers that
hundreds of thousands of jobs would be lost, thereby derailing the
accelerating recovery in employment that had been taking place, and
harming the economy further in the long run.

He blamed the wrong turn on ideology and politics, and saw the
highway problem as merely an example of a more widespread problem.
He reiterated the usual political and economic reasons why the gas
tax—and the price of fuel—was being kept artificially low and still
argued for increasing it, citing issues ranging from climate concerns to
dependence on Middle East oil. He recognized that political realities in
Washington might argue even more strongly against increasing the tax,
but, he emphasized, even without that source of revenue to keep the
trust fund solvent, as a nation "we don't have to stop building and
repairing roads."

In a paragraph characterizing deficits and taxes as "evil" and thus to
be avoided, Krugman concluded sarcastically that "our roads must be
allowed to fall into disrepair." He blamed similarly "crazy" logic for the
nation's "overall plunge in public investment," most of which is done by
state and local governments, most of which are constrained to balance
their budget, most of which are short on revenue themselves. But,
argued Krugman, since the federal government is not so constrained, it
could support public sector investment by means of grants that were
deficit financed. Furthermore, the states themselves could increase their
revenue stream through increased taxes and other means, but mostly
did not do so for reasons similar to those of Washington. So, Krugman
concluded, "the collapse of public investment was, therefore, a political
choice." This "self-destructive" choice "means that we're letting our
highways, and our future, erode away."

This was not a new realization. Long before the federal government
got into the infrastructure-funding business, the states made such

investments. Early in the nineteenth century, New York issued bonds to finance the construction of the Erie Canal, which opened in 1825 and was hailed as a great success both technologically and economically. Other state governments followed New York's example and very quickly began to rack up large debts. A national total of $26 million was borrowed in the decade of the 1820s, $40 million in the half decade of the early 1830s, and more than $100 million in just the three years of 1836 to 1838.

Pennsylvania, which was New York's principal rival for controlling inland trade, was the largest borrower of all the states. It went into debt to build a system of canals, railroads, and roads between the Delaware River at the port of Philadelphia and the beginning of the Ohio at Pittsburgh, which was a gateway to the West. By the end of 1841 the state of Pennsylvania owed about $40 million, which meant $2 million in interest annually. At the time, the state's annual revenues to cover the debt were not quite $1 million. It raised taxes but still could not cover its debt. It tried other schemes, including forcing banks with state charters to lend money to cover the deficit. Nothing was enough, and in 1842 Pennsylvania defaulted. Several states defaulted during the same period of deflation. Another wave of defaults, which occurred mostly in southern states following the Civil War, has been blamed largely on carpetbaggers. A few states defaulted on bonds during the Great Depression, but since then state-incurred obligations have been relatively safe investments. Of course, not all questions put to voters are approved, and so some states have had difficulty in funding public works projects.

In 2014, almost 59 percent of Missouri voters rejected a constitutional amendment to increase the state sales tax by three-quarters of a percent in order to raise $5.4 billion over ten years for a long list of road, bridge, and other infrastructure projects. The increase, when added to local sales taxes, would have resulted in an overall sales tax as high as 10 or 11 percent in some areas of the state. Critics of the amendment called the tax regressive and believed an increased tax on fuel would have been preferred. Missouri Department of Transportation officials argued that the fuel tax had not been supporting the state's highway maintenance budget, let alone new construction, to the extent desired. The disappointed sponsor of the amendment said after the vote, "When people start seeing bridges close, they're going to say, 'Hey, we do have a problem.'" But for every Missouri there may be a Texas. In the 2014 midterm elections, 81 percent of Texas voters gave the state the go-ahead to put half of oil- and gas-field revenues into a highway fund, but stipulated that it could not be used for toll roads. The vote may have reflected Texans' strong distaste for toll roads at least as much as it did their desire to fund highway programs.

The results of a nationwide poll reported late in 2014 showed that although a majority of voters were in favor of increased funding to repair roads and bridges, two out of three U.S. residents were opposed to increasing the federal gas tax to pay for transportation projects. Thus, the states were challenged to be creative. Washington State looked at imposing a carbon charge on industrial polluters. Over twelve years, the fee was expected to generate about $7 billion, roughly as much revenue as a 12-cent-per-gallon increase in the gas tax. Governor Jay Inslee considered the approach "a pretty elegant solution" for his state. North Carolina governor Pat McCrory was among state chief executives looking at revenue shortfalls of the order of tens of billions of dollars over the next generation, but on the eve of the 2015 legislative session he was reported to be keeping close to the vest his plans for closing the gap. His twenty-five-year vision for the state recognized that "transportation infrastructure plays a critical role in attracting and retaining businesses" and so his "Mapping Our Future" initiative focused on how to "leverage our infrastructure to catalyze economic growth" while at the same time recognizing that different regions of the state had different needs.

Given the citizenry's support for infrastructure investment but not necessarily for its coming out of their pocket, we can expect to see increasingly more creative solutions being proposed in the coming years. However, fixing the holes in voter and political logic will continue to be a far more difficult problem to solve than fixing potholes. Krugman's logical analysis of illogical Washington gives us a clear linear explanation of why we are where we are in terms of infrastructure, but it leaves a lot unsaid.

Since 1998 the American Society of Civil Engineers' periodic report cards on the condition of the nation's infrastructure have estimated how much money is needed to bring things up to acceptable standards. Following the 2009 report card, the organization issued a series of reports under the title *Failure to Act*, which assessed how the near-failing state of the infrastructure was affecting the economic performance of the nation. In particular, *Failure to Act* analyzed the economic implications of simply continuing then-current trends in infrastructure investment in a variety of categories, including those of roads and bridges and other forms of surface transportation, with regard to building new and maintaining existing ones.

Among the reasons that inadequate and poorly maintained infrastructure impacts the economy negatively are that the cost to operate a car or truck increases, because time spent in traffic congestion increases transit time and fuel consumption—both of which can be converted into dollar costs—and because roads in poor condition are harder on vehicle suspension and other systems, which translates to higher-than-normal maintenance and repair costs. Such effects further impact the overall economy by increasing the cost of shipping goods and delivering services, and they also reduce the amount of disposable income individuals have to spend on things other than vehicle operation and maintenance. The ASCE reports estimated that nearly nine hundred thousand jobs were expected to be lost by 2020 due to the continued economic impacts of a poor transportation infrastructure alone. However, by 2040, this number is projected to drop to about four hundred thousand because of expansion of industries associated with automotive repair. Such a "benefit" does not deserve to be called a silver lining.

The bottom line of the *Failure to Act* report series is that cumulative overall infrastructure needs (in 2010 dollars) by 2020 will be underfunded by $1.1 trillion and by 2040 by $4.7 trillion. For surface transportation needs alone, the totals are $846 billion and $3.7 trillion, respectively, with improving "pavement and bridge conditions" responsible for a good proportion of the total. According to the ASCE, an investment of $157 billion per year across all sectors of infrastructure through the year 2020 will prevent the loss of $3.3 trillion in gross domestic product and 3.5 million jobs. From where can such an investment come? And what do we do with infrastructure in the meantime?

19

Ever Come Back

PEDESTRIANS, PRESERVATIONISTS, PARKS

INFRASTRUCTURE INVESTMENT MAY BE thought of as a down payment on the future. As such, there is often little enthusiasm for throwing good money after bad to revitalize something old, obsolete, unused, unproductive, unsightly, and unpromising. Whether it is an abandoned factory or a dried-up canal, a dilapidated piece of infrastructure is seldom seen as something to be restored when proposals for new businesses employing modern tools in modern buildings to make modern products are begging for investment to help an old mill town become revitalized. The old may have architectural charm and historic significance that deserves to be preserved as part of our heritage, but new structures promise opportunity, jobs, and hope for a better future. Can a struggling economy really adequately both tend to the old and nourish the new?

A bridge is typically built for a specific purpose. Hence we have railroad bridges, highway bridges, pedestrian bridges, pipeline bridges. Some bridges serve multiple purposes. No matter the original intent, the use of any bridge may change over time for a variety of reasons, ranging from the changing ways and means of the communities and industries that it serves to the deterioration or obsolescence of the structure itself. Many a historically important structure was designed to carry traffic that was different in character, weight, and volume than what crosses it in its advanced age. Trucks are much heavier now than they were even fifty years ago. Some bridges today are being called

upon to carry several times as many cars and trucks as they were
designed to, and so are deteriorating at a faster rate. If neglect of proper
maintenance is added to this, a steel bridge can corrode, losing its
strength and accelerating its decline.

When traffic overwhelms an aging and deteriorating bridge, a
replacement is often constructed beside it or nearby in order to connect
to existing roads, as has been done with the east span of San Francisco's
Bay Bridge and is being done with New York State's Tappan Zee. What
to do with the old span once its traffic load has been transferred to the
new is a problem that communities and state departments of transportation are facing with increasing frequency. If the old structure is left in
place, perhaps to carry the lighter load of local traffic only, it must still
be maintained. If the old structure is demolished, the money spent to do
so cannot be available for what might appear to be more pressing
projects, such as repaving roads full of potholes or building a new road
or bridge. The decision is further complicated for historically note-
worthy structures. Their very significance usually brings forth considerable support for preserving them in, if not restoring them to, their
original form, which can be a very expensive proposition.

Among the most popular approaches to preserving historically significant bridges today is to convert them to pedestrian and bicycle use,
often incorporating them into a dedicated walking and cycling trail or a
park. An especially ambitious approach to conversion was taken with a
long-disused railroad bridge over the Hudson River at Poughkeepsie,
New York. A park and trail were built leading up to and across the long,
high structure as part of a project known as Walkway Over the Hudson.
The path to achieving this, like that leading up to the completion of the
bridge itself, was arduous and spanned decades.

SOME OF THE EARLIEST serious discussions of building the first bridge
across the Hudson River south of Albany took place in the 1860s. At
that time government was not expected to undertake such a project, and
so private enterprise assumed the responsibility—and the risk. A fixed
crossing of the Hudson located somewhat north of New York City

would minimize the travel distance to Boston for trains coming from the anthracite fields of Pennsylvania, and so coal companies could be expected to promote use of the bridge and contribute to a revenue stream that would provide a return on investment. Such a fixed link would also facilitate commercial and passenger rail traffic between New England generally and states to the west and south. Among the proponents of a bridge was the Hudson Highland Suspension Bridge Company, which engaged engineers to evaluate the idea of erecting a span just north of the town of Peekskill, located at a narrow part of the river about forty miles up from its mouth. The engineers identified a suitable location where a bridge could join equally elevated sites on the west and east banks, essentially where the Bear Mountain suspension bridge would eventually be built.

In the meantime, a bridge was also being promoted in a larger town about thirty miles upriver. Poughkeepsie is located on the east bank of the Hudson about midway between Albany and New York City. The idea of a bridge at Poughkeepsie actually predated that for one at Peekskill, and soon became the leading contender. Legislation passed in 1871 granted a charter to a bridge company incorporated by, among others, Matthew Vassar, founder of the nearby academy for women named after him. Unfortunately, a clause in the act prohibited locating any bridge piers in the river itself, a constraint that created a challenging engineering problem. A suspension bridge might have required a clear span of 2,600 feet, which was well beyond the state of the art. When this was communicated to the state legislature, an amended charter allowed for four piers to be located in the river, making each of the individual spans some 500 feet long. This would provide plenty of room for passage of the canal boats that were lashed together to form a "tug" as large as 150 feet wide and 800 feet long. The Eads Bridge under construction across the Mississippi at St. Louis at the time provided precedent for the 500-foot spans, and its chief engineer, James B. Eads, was frequently invoked in making the case for a Poughkeepsie cantilever bridge. The charter required that construction begin before 1874, and when investors associated with the Pennsylvania Railroad committed more than $1 million to the project, that became possible.

John O'Rourke, who would serve as resident chief engineer for the Union Bridge Company, which would eventually design and build the structure, reported that two weeks before the 1874 deadline, amid "elaborate Masonic ceremonies and general rejoicing," a cornerstone was laid for one of the piers. Actual work was begun in 1876 by American Bridge Company, but a financial depression soon caused support from investors to dry up, and not until 1886 did full-scale work commence under the Union Bridge Company.

During the academic year 1887–1888, Thomas C. Clarke, who was involved not only with the foundations of the bridge but also with general bridge-building operations, gave a lecture at Cornell University on the still-incomplete structure. He began by noting that "modern bridge building is the creature of the railway system, and its productions are as different from the stately stone arches of our ancestors as is a locomotive from a Greek chariot." He estimated that, of the 150,000 miles of railways then in the United States, there was "at least one bridge of 100 ft. long to every mile." (In this regard he echoed what the innovative bridge engineer Thomas Moseley had written two decades earlier: "Almost every individual who, as its engineer, has made ten miles of road, has at one time or another conceived a new plan of bridge; for of all the troubles which beset an engineer in constructing and operating a road, it's Bridging is the greatest.") Clarke further noted that because so many bridges had to be built in a relatively short period of time, there was little regard for aesthetics and that "utility alone governs their design." He thus prepared his audience for the lantern slides of the bridge that he admitted was not beautiful but that he hoped would bring joy to stockholders. Clarke assured his audience that the "considerable piece of engineering" that he was showing them was "one of the great bridges of the world."

The Poughkeepsie railroad bridge was completed by the end of 1888, but its tracks were not yet connected to a single established railway. In order for a ceremonial first train to reach the bridge and cross it, temporary tracks had to be laid. But since there were no connections on the other side, either, the train could do little but shuttle back and forth atop the great structural achievement, and another six months passed

The Poughkeepsie Railroad Bridge, completed in 1888, was built to enable trains to bring anthracite coal directly from western Pennsylvania to New England. With the decline in the use of coal and of the railroads generally, the high bridge became a white elephant. Demolition would have been complicated and expensive, but visionaries repurposed the structure to make its spectacular views of the Hudson river valley available to everyone. Today it is the centerpiece of a unique pedestrian experience known as Walkway Over the Hudson.

before the bridge was indeed connected to regional railroads. The Poughkeepsie-Highland Railroad Bridge, as it was sometimes called—Highland referring to the unincorporated hamlet across the river in Ulster County—would never be a great commercial success. Still, it had to be maintained and upgraded if there was even to be a hope of its lasting long enough to be profitable. In the meantime, railroad rolling stock and their loads were continuing to grow heavier, and so in the early twentieth century the bridge was strengthened by the costly addition of a third vertical truss down the middle of the structure.

Talk of pedestrian access to the Poughkeepsie Railroad Bridge had begun even before it was completed. A local paper called for allowing ordinary citizens to pay a toll for access to the narrow inspection and maintenance walkways of the high bridge as an alternative to relying on a ferry to cross the river. The bridge charter was even amended to allow for pedestrians. However, the bridge company balked, expressing concern about the fire danger people might pose by smoking, not to mention the dangers crossing trains would present to people walking beside the tracks. The owners continued to bar pedestrian access even though the bridge was considered to provide "perhaps the most restful, placid, and pleasing river view in the world." Local activists mounted challenges now and then to pedestrian exclusion, but the countervailing issues of safety and the cost of insurance to the bridge owners prevailed.

In the 1920s there was renewed interest in being able to walk across the river at Poughkeepsie. With the growth of automobile traffic, there were calls also for a roadway on the bridge. The septuagenarian engineer Gustav Lindenthal was asked his advice on what would need to be done to the structure to allow for the desired pedestrian and vehicular traffic. His conclusion was that the amount of work that would be needed was economically prohibitive, and so nothing was done to augment the railroad function. Eventually the bridge came to be owned by the New Haven Railroad, which, with a lot of other later twentieth-century railroads, would in time go bankrupt.

The interest in a pedestrian and vehicular crossing of the Hudson at Poughkeepsie led to the development of plans for a second bridge, and the Hudson Valley Bridge Association was formed to promote a new span. Whereas the economic argument for a Poughkeepsie railroad bridge had been the transport of coal and manufactured goods across the Hudson, the argument in the fledgling age of trucks was the transport of farm goods. To bolster its contention that a bridge was feasible, the association engaged George W. Goethals, whose accomplishment in overseeing the completion of the Panama Canal had made him legendary. He had not done much bridge designing in his engineering career, however; the proposal he came up with for Poughkeepsie was another cantilever, which differed in configuration from the railroad structure mainly in having the roadway contained within the trusses rather than atop them.

Nevertheless, Goethals' design appeared to be sufficiently sound to convince Frederick S. Greene, the chief engineer of the state's Department of Public Works (and, incidentally, a horror fiction writer), to go ahead with a new bridge. However, it would be designed not by Goethals but by Ralph Modjeski, whose bridge experience was broad and deep. It was Modjeski who had designed the reinforcement of the Poughkeepsie railroad bridge by means of its third truss, and so he was familiar with the geographical location and the problems it posed for construction. Acutely aware of the challenges presented by the river itself, he went into partnership with Daniel E. Moran, whose specialty was foundations. Modjeski and Moran proposed a suspension bridge whose towers rested on piers located 750 feet out from each riverbank and whose cables supported a 1,500-foot center span 135 feet above the water. The Mid-Hudson Bridge was completed in 1930, its roadway flanked by four-and-a-half-foot-wide sidewalks. However, access to the sidewalks proved difficult: the speeding age of the automobile had distracted attention from such pedestrian concerns.

With the establishment of the trucking industry in the 1930s and its substantial growth in the late 1950s and early 1960s, which coincided with the rise of the interstate highway system and the concomitant decline of the railroads due in part to the relocation of American heavy industry away from the Northeast, bridges like the Poughkeepsie-Highland carried fewer and fewer trains and eventually fell into neglect and disrepair. The turning point came around 1970, about the time when the Penn Central Railroad took over control of the bridge. As a cost-saving measure, the number of workers assigned to maintaining and guarding the bridge was cut, which increased the risk that a fire started by a spark from a passing train would not be spotted before it did serious damage. This happened in 1974 when a fire was ignited in the timber ties, which burned until some of the steel below was damaged by the heat. The cost to restore the bridge was estimated to be $650,000, which the bankrupt Penn Central could not afford.

In the wake of the fire, there were discussions among Penn Central, the state of New York, and the U.S. Department of Transportation about how to finance needed repairs, but the parties could not come to an agreement. In 1976, Penn Central was taken over by the newly created quasi-public Consolidated Rail Corporation (Conrail), which wished that the liability that the Poughkeepsie Bridge had become did not exist. By 1980, rehabilitation of the bridge was estimated to cost almost $20 million. Conrail looked to tearing it down, but proposals to blow up the structure and retrieve the parts from the water were met with opposition. Not only would such an operation interfere with boat traffic but it also would stir up the river bottom, which was heavily contaminated with polychlorinated biphenyls, substances that Schenectady-based General Electric had used as insulation coatings in the manufacture of electrical devices.

In the meantime, a number of proposals emerged for preserving the high bridge by converting it to a new use. Architect Edmond Loedy of nearby Millbrook, New York, proposed incorporating the bridge into a linear mall containing shops, restaurants, hotels, condominiums, and observation towers. Robert Turner, an architecture student, proposed running a light-rail line across the bridge, which would also have included paths for jogging, bicycling, and sightseeing. At each end of the bridge, office towers incorporating museums and restaurants were planned. In 1984, Conrail sold the bridge to Gordon Miller, an elusive absentee private investor, who not only neglected to maintain it and pay taxes on it but also endangered the structure and boats on the river by not illuminating its navigation lights. In 1990, a bankrupt Miller sold the bridge for one dollar to Vito Moreno, said to be an electronics engineer from a Philadelphia suburb, who proved to be equally frustrating to deal with for the Poughkeepsie community. Moreno did allow access to two bungee jumpers from Colorado, who proposed basing a bungee-jumping business on the bridge, but the local community was not supportive, some thinking that it would "detract from the bridge's dignity."

In 1992, the not-for-profit organization Walkway Over the Hudson was formed with the stated mission to "preserve the landmark

Poughkeepsie-Highland railroad bridge, transform the bridge into a linear park and trailway and to provide long-term stewardship." The high bridge parkway would provide spectacular views of the scenic Hudson Valley and, through connecting railbed trails, provide pedestrian access to the parks and communities in the area. In 1993, members of Walkway Over the Hudson claimed ownership of the Poughkeepsie Bridge by squatting on it, asserting that since Moreno did not pay taxes or fines from the Coast Guard, he had effectively abandoned the structure. The following year, volunteers made the western part of the bridge suitable for walking upon, and the spectacular views helped generate influential support for its preservation. In 1998, Moreno abandoned any hope of developing the bridge for profit and gave up his claim to it to Walkway Over the Hudson.

In 1999, with reconfigured approaches, the sidewalks on the Mid-Hudson Bridge (since 1994 officially named the Franklin D. Roosevelt Mid-Hudson Bridge, in honor of the former governor of New York who had lived nearby) were made more accessible not only to outdoors people but also to people with disabilities, and in the meantime the unused and neglected railroad bridge just a mile upriver was increasingly eyed as an underappreciated resource. Walkway Over the Hudson eventually partnered with the Dyson Foundation, a local philanthropy that provided a grant to assess the condition of the structure and to raise funds to "transform the bridge into the world's largest pedestrian park." Indeed, the rehabilitation of the bridge was to be "for re-use as an iconic New York State Park." Groundbreaking, in which some remaining rails were symbolically removed, took place in 2008 and soon construction "creating the world's longest pedestrian bridge" began.

Thirty years earlier, the structure had been nominated for the National Register of Historic Places under the augmented name of the Great Poughkeepsie Railroad Bridge, and it had been added to the register as the Poughkeepsie Railroad Bridge the following year. In the application, the total length of the bridge was listed as 6,767 feet (including approach viaducts) and it was classified as having restricted accessibility. In 2008, the structure was nominated for designation by the American Society of Civil Engineers as a National Historic Civil

Engineering Landmark, and a plaque identifying it as such was unveiled on October 17, 2009, a week after the formal dedication of Walkway Over the Hudson. Not coincidentally, it was the year marking the four hundredth anniversary of Henry Hudson's exploratory voyage up the river that now bears his name. In the first eight weeks the walkway was open it attracted four hundred thousand visitors.

THE STORY OF THE recycling of the Poughkeepsie bridge is only one of many that can be told of similar projects across the country. In Manchester, New Hampshire, an abandoned railroad bridge was rehabilitated and reopened in 2008 as Hands Across the Merrimack, which enables pedestrians and bicyclists to move freely and pleasantly between the east and west sides of the city. The Big Four Bridge, which once carried the Cleveland, Cincinnati, Chicago & St. Louis Railway (the "Big Four Railroad") across the Ohio River between Louisville, Kentucky, and Jeffersonville, Indiana—had been abandoned for decades. After public access ways were removed in 1969, it became known as the "Bridge That Goes Nowhere." The long-disused bridge was renovated by the Louisville Waterfront Development Corporation, with the Indiana approach overseen by the city of Jeffersonville. Bad economic times threatened the project, but in 2011 the states of Kentucky and Indiana and the city of Jeffersonville allocated $22 million to complete it. In 2013, access to the bridge via a helical ramp in Louisville's Waterfront Park was opened, and the following year an access ramp was completed on the Indiana side. Thus another historic railway bridge became available to pedestrians and riders of nonmotorized vehicles.

Still another, known as the High Line, was a long-disused elevated railway spur that runs parallel to the Hudson River along Manhattan's West Side. Historically, it had served to transport railroad cars carrying cattle and industrial goods into and out of New York City but had not been in use since 1980. Calls for its demolition were met with pleas for its preservation and renovation. The ardor, perseverance, and support of numerous New Yorkers, drawn from the ranks of writers, artists, designers, media people, Wall Street managers, politicians, and others, paid off when the first phase of an elevated park opened in mid-2009. Creatively designed wooden walkways stretched for about ten city blocks beside railroad tracks that were left in place. From between boards, rails, and ties grow more than a hundred species of plants representing the wild landscape that had continued to take root when the trains ceased running. Many of the wild seeds were no doubt hitch-hikers on the railroad cars that came from all parts of the country to ride above the asphalt and concrete below. The third and final section of the High Line, extending the park itself another dozen or so blocks farther north, was completed in 2014. The next year, the Whitney Museum of American Art relocated from its landmark Marcel Breuer building on the Upper East Side into a larger and new Renzo Piano–designed building at the southernmost end of the High Line, in the city's old meatpacking district, which was becoming revitalized. The iconic Breuer building itself was repurposed by being leased to the Metropolitan Museum of Art as expansion exhibition space.

Such achievements as these are encouraging models for recycling historic parts of our infrastructure into modern recreational and cultural use. They show that the road to completion may be a long and rocky one, but that the final results can be gratifying indeed. The money saved by not demolishing old bridges and buildings can be spent instead on different infrastructure and preservation projects.

20

Telling This with a Sigh

PUBLIC-PRIVATE PARTNERSHIPS: PLUSES AND MINUSES

FINANCIAL SUPPORT FOR ENVIRONMENTALLY friendly infrastructure reclamation projects like Walkway Over the Hudson and the New York City High Line may be able to be raised by well-connected and persuasive organizers from sympathetic friends, acquaintances, and foundations. However, generating sufficient funds or even just moral support for a new highway, bridge, or other public works project—let alone for its maintenance—can be a long and lonely road to disappointment. We have come to expect government agencies to do the groundwork on such infrastructure needs, and municipal councils and state and federal legislatures to do the authorizing and appropriating of the necessary funding. But, especially in times of scarce resources and fiscal belt-tightening, government agencies recognize the public's desire for many more projects than can reasonably be supported.

For the last century or so in the United States, the generally expected route to having an infrastructure project such as a major bridge approved and funded has been for a state department of transportation to give the project a high priority and then seek the funding either in its own budget or from the federal government. Before that, it had been much more common for the U.S. government to stay out of the funding business altogether, but a notable nineteenth-century example of government assistance occurred with federal bond issues and land grants associated with the completion of the first transcontinental railroad. Roads, bridges, railways, and the like were seen by entrepreneurs and corporations as attractive investments that would over time be repaid with interest. The needed capital outlay, which admittedly could be considerable, would be recouped in revenue from fares, tolls, and other user fees. Obviously, an accurate prediction of future traffic volume and economic developments would be essential to determining whether investments would eventually pay off.

A legislative charter concession, or franchise assured a bridge-building company and its investors that they would have no competition for a specified length of time. The arrangement was not unlike the patent system, whereby inventors are granted a monopoly for a specified period in exchange for revealing the details of their invention. As the U.S. Constitution states in Article I, Section 8, this was done "to promote the Progress of Science and useful Arts," meaning that there was a quid pro quo. Had an inventor not received the guarantee of an exclusive right to his invention, once introduced into the marketplace it could be copied and sold at a lower price by a competitor who had not invested any fiscal or intellectual capital in the invention's development. Under the patent system, the nation gains by having the inventor's thinking about the device or process laid out in a declaration for other inventors to study and improve upon, thereby adding further to the public store of useful knowledge and things that came of it. While many innovative bridge-building details and procedures are patented, generally speaking, the workings of a completed bridge are open for all to see and so improve upon.

The twentieth century brought a general shift from private to public investment in infrastructure projects. One of the last great projects privately funded was the Ambassador Bridge connecting Detroit and Windsor, Ontario, still operating as a private enterprise almost a century later. Talk of a bridge or tunnel increased in the 1880s when it became evident that ferry service across the Detroit River was being stretched, especially by the number of railroad cars and by winter ice conditions. A number or proposals were put forth, including one by the highly regarded bridge engineer Gustav Lindenthal, but its price tag of $6 million was considered prohibitive, even for the railroads. Despite federal government support that should have made it relatively easy to

secure a concession for an international crossing, local political
disagreements, conservative business practices, and shipping interests
continued to obstruct progress toward one. No fixed crossing came to
be until a railroad tunnel was opened in 1910.

In the meantime, with the development of the automobile industry
in the Detroit-Windsor area and an increasing number of motor vehicles
on the road, there were renewed calls for a bridge. Even businessmen
had changed from opponents to proponents of the idea. Engineer
Lindenthal, who had been promoting long-span bridge designs to cross
the Hudson River into New York City, again became involved in the
Detroit River challenge. He teamed up with the younger engineer
Charles Evan Fowler, who had recently proposed a massive cantilever
bridge connecting Oakland and San Francisco across the bay. The two
engineers could not agree on a design for a Detroit River crossing,
however, and the aging Lindenthal withdrew from the project, perhaps
to devote more time to his ever-grander scheme to cross the Hudson
River.

Fowler came up with a suspension-bridge design that would carry
highway traffic on an upper deck and railroad trains on a lower one.
The price tag was estimated at $28 million. Eliminating the railroad
capacity lowered the price to about $11.5 million. In order to finance the
project, in April 1920 Fowler formed the American Transit Company; a
year later he formed the Canadian Transit Company. Each was established with the approval of the respective government, and the companies were to jointly finance and build the international crossing. After
producing detailed plans, which were reviewed and endorsed by a distinguished board of consulting engineers, Fowler sought the support
of business and community leaders as a prelude to seeking financial
support. He had expected the railroads to help financially, but they
already had a tunnel crossing and so expressed little interest in the
bridge.

The railroad deck was eliminated, and Fowler now promoted a
pared-down bridge design. All the needed government approvals were
received in 1921. Federal legislation passed in Washington and Ottawa
stipulated that construction of the bridge had to begin within three

years of the bills becoming law and the structure had to be completed within seven years. Such was the nature of a concession.

Fowler saw the enterprise as one undertaken for the public good, and as such the bridge was to be "built and financed solely for the public benefit without profits accruing to the promoters further than a reasonable compensation to the directors for their three to four years tenure of office." Stock was to be sold and bonds issued to raise the needed money, but a fund-raising partner seemed not to be as public-spirited as Fowler, and unethical behavior and scandal gave the project a bad name. When it became clear that there was not sufficient support for a privately owned and operated international bridge, Fowler spent more and more of his time on other engineering projects.

In 1923, James W. Austin, one of Fowler's Detroit associates who had been treasurer of the American and Canadian transit companies, took the lead in promoting the bridge. He soon realized that there was not enough investment capital in the Detroit and Windsor area to fund the project and reached out to the large Pittsburgh-based steel-fabricating company McClintic-Marshall. The company was interested in building the bridge but could not afford to finance it. C. D. Marshall, one of the company's partners, introduced Austin to New York financier Joseph A. Bower, a onetime Detroiter working in New York banking. Once he had become convinced of the soundness of the bridge proposal as a technological endeavor and an investment opportunity, Bower bought out the transit companies, paid off their liabilities, and engaged Fowler as consulting engineer. Soon thereafter, Bower declared that he would finance the bridge through "a company of a few friends" who put in a million dollars. The balance would be raised through bond issues.

Construction of the bridge was begun on May 7, 1927, just six days before the expiration date of its congressional authorization and under a contract that imposed severe penalties for not meeting a three-year construction schedule: McClintic-Marshall would have to pay all the projected tolls that would not be realized because of late completion; to balance the penalty, the company would share equally in toll revenue between the time of an early completion and the scheduled opening.

WINDSOR, ONTARIO   AMBASSADOR BRIDGE   DETROIT, MICHIGAN

This postcard view of the privately financed, built, and owned Ambassador Bridge looks across the Detroit River from Windsor, Ontario, on the Canadian side. When the suspension bridge opened in 1929, its 1,850-foot main span was the longest of any bridge in the world. While fully private infrastructure projects have become rare, a scarcity of government funds has increased interest in public-private partnerships for building major roads and bridges.

When the privately funded, owned, and operated international suspension bridge did open in 1929, it boasted the longest span in the world: 1,850 feet between its tall steel towers, each of which four years later would be topped with a large sign reading, AMBASSADOR BRIDGE. Bower had thought the expected name "Detroit-Windsor International Bridge" was "too long and lacked emotional appeal," and he had rejected urgings to name it "Bower Bridge," thinking that improper. Of the name decided upon before the structure was completed, he had explained, "I thought of the bridge as an ambassador between two countries, so that's what I called it."

The Ambassador Bridge opened to traffic a year earlier than the mile-long Detroit–Windsor Tunnel owned equally by the cities of Detroit and Windsor. With the tunnel as an option, traffic on the bridge understandably dropped, and consequently so did toll revenue. The Great

Depression brought a decline in road traffic generally, and soon both bridge and tunnel were defaulting on their debts. The bridge company was reorganized in the late 1930s, but World War II brought further-decreased revenue and more tough times financially. Not until the postwar years did prosperity return to the country and the bridge. The crossing continued to be owned by the Bower family until the patriarch's death in 1979, when it was bought by another family-owned business, the Central Cartage Company of Detroit, headed by Manuel J. Moroun.

In 1988, with the signing of a free-trade agreement between the United States and Canada, the bridge became notorious for heavy, slow-moving truck traffic, which had to use local streets in Windsor. Upon the completion in 2012 of ramp connections providing a direct route between bridge and highways, Matthew Moroun, vice-chairman of the bridge company and son of its head, released a statement that read in part, "We will not thank the government for their efforts to damage our business during this project." The animosity between the private bridge company and the government stemmed from an ongoing disagreement about how to supplement the Ambassador Bridge to achieve greater traffic capacity over the U.S.-Canadian border.

In the meantime a new competitor loomed on the horizon. The New International Trade Crossing project, which dated from 2004, was to be jointly paid for by Canada and Michigan, the latter anticipating funding from the U.S. Federal Highway Administration. There was strong Canadian support for the project, but Michigan politics did not cooperate, and so planning dragged on. Nevertheless, in time a Canadian crown corporation—essentially a partnership between public and private entities—was established; it was known as the Windsor-Detroit Bridge Authority. In late summer 2014 the authority had one employee, its CEO Michael Cautillo, an Italian-Canadian civil engineer who came from a career of work on transportation infrastructure, most of it with the Ontario Ministry of Transportation. His broad experience in that position was expected to serve him well, since the bridge project would involve not only the major crossing of the Detroit River but also smaller bridges connecting it up to a Canadian expressway and an American interstate highway, not to mention the ports of entry and a toll plaza. He expected the construction project to go out for bids sometime in 2015. In the meantime, he would build up an authority staff of as many as fifty.

Not surprisingly, the owners of the Ambassador Bridge—which in early 2015 carried more than 25 percent of all U.S.–Canadian commerce—had voiced strong opposition to the new government-sponsored crossing. Matthew Moroun accused the government part-ners of using "political math" to justify a $2 billion project. According to Moroun, traffic volume on the Ambassador Bridge was bringing in through tolls and other revenue an annual total of about $60 million, out of which had to be paid operation and maintenance costs. What was left could hardly justify investing $2 billion in a new bridge that would have to attract traffic from the Ambassador Bridge and other area crossings. Furthermore, competition would argue for lowering rather than raising tolls. By Moroun's math, the government enterprise would need $80 million per year just to pay off the interest on its bonds. The taxpayer would have to make up the difference.

Moroun's company had its own plan for expansion: a new bridge right beside the Ambassador, which had needed work even before it reached seventy-five years of age in 2004. The Ambassador Bridge Enhancement Project called for a $1 billion private investment in a cable-stayed bridge that would tie into the same approach roads and so require minimal additional new infrastructure. When the new bridge opened to traffic, the old Ambassador would be closed for extensive work. Being able to renovate the bridge without traffic on it would cost about one-quarter what it otherwise would. The renovated old bridge would be used for maintenance vehicles, for overflow traffic on the new span, and for "special events." One can imagine those being revenue producing.

EVEN AS THE AMBASSADOR Bridge conflict was providing an ongoing cautionary tale demonstrating that public and private interests do not always see eye to eye, an old model of how local and state governments could finance large American infrastructure projects was being revived. The idea was to create public-private partnerships, the mouthful often being abbreviated PPP or P3 by participants, analysts, proponents, and critics alike. The concept was not new to Europe or Canada, the famous Hudson's Bay Company having been established in 1670 on that principle. In the United States, the concept goes back at least to 1785, when the first U.S. toll bridge was opened, connecting Boston and Charlestown across the Charles River. According to the legislative charter granted to the Charles River Bridge Company, it would fund and construct the bridge and for forty years keep all the tolls. At the end of that period, the bridge would become the property of the Commonwealth of Massachusetts. It was such a successful business venture—with reported annual rates of return to investors being as high as 40 percent—that similarly chartered toll bridges multiplied.

Under a public-private partnership, what have generally come to be thought of as public works—highways, bridges, parking garages, waterworks, and even government buildings—are being financed, designed, built, operated, and maintained by for-profit corporations for, usually, a multi-decades-long period of time. In exchange for the private money invested in a project, the corporation typically gains the exclusive right to collect tolls, fees, rents, and the like for the duration of the agreement. At the end of the contract, the facility is turned over to a public entity or a new contract negotiated. If done right, both partners see themselves getting a good deal.

A public-private partnership is distinctly different from the concept of privatization, which was what the British government did in the mid-1990s when it broke up the national railway system British Rail and sold the pieces to private companies. That action was prompted by the beliefs of the government in power at the time: "The Tories felt the railroad's monopoly status encouraged bureaucracy, low productivity, and inattentiveness to customer needs. The government believed the antidote was markets and competition, which would promote efficiency and innovation." Perhaps mostly because the change was made hastily, with many concessions given to the purchasers, privatization turned out to be a disaster. But renationalization has been said to be even worse.

In a public-private partnership, no public property is sold. The public partner, a city or state or government agency, typically receives a large payment up front for granting the private partner a concession: the use of land on which to construct the project, or the use of existing facilities to manage. The up-front payment, perhaps billions of dollars, can fund other infrastructure projects or shore up a neglected pension fund. The private partner is willing to make such a large payment because it is far exceeded by projections of revenue throughout the lifetime of the agreement, which might run for fifty years or more. The calculation of projected revenue over initial capital investment is what attracts investors, who benefit from a favorable long-term cash flow. Since multiple billions of dollars are typically involved, large investors like investment banks, state teacher retirement programs, and other city and state pension funds are seen as prime targets for the investment opportunity. From this perspective, a P3 arrangement looks like a win-win deal.

But a partnership of any kind comes with risk. In the case of a P3, projections of future use, and hence of future revenue, can be overly optimistic, in which case the operating partner will wish to raise tolls, fees, or rents to compensate. Users of the facility will of course not like this, and they will seek alternatives, thus lowering use even further. A toll highway may be a public-private partnership, but to users and voters it will look like a public work, and so the nearest politicians will likely be blamed for increased tolls. The subtleties of ownership and control will be lost on the general citizenry, who will vote by anecdote rather than by fact. A disappointed private partner may skimp on maintenance to save money, thereby presenting the appearance of neglect that will also likely be laid at the feet of politicians. A win-win situation could easily turn into a lose-lose one.

All contingencies should, of course, be anticipated by both parties, and a carefully drawn contract between them should protect each from the other. Unfortunately, for a variety of reasons, achieving a fair and balanced contract—especially for the public party—is not easy. Each side in the negotiations wishes to have an advantage, and the more experienced and forceful negotiators tend to be on the private party's side. Generally, government bureaucrats are not likely to have had much if any experience with such nonstandard contracts, and so are at an immediate disadvantage. The government entity itself may not be able to afford consultants and attorneys well suited to represent the public interest. The private interests, on the other hand, are typically well represented at the negotiating table, because they have retained counselors precisely for their experience and tough negotiating stance—and investors risking billions of dollars are willing to pay top dollar for top-notch advice. Nevertheless, in spite of the risks to the public interests, by mid-2014 more than thirty states had passed legislation enabling P3 arrangements for transportation projects. And although the widespread use of the public-private partnership model is still relatively young in the United States, a number of instructive case studies exist from which lessons can be drawn.

In 2009, Chicago was facing a budget deficit, which then mayor Richard M. Daley solved by striking a deal with a private venture organized by the international financial services corporation Morgan Stanley. In exchange for control of thirty-six thousand of the city's metered parking spaces, Chicago would receive an initial payment of $1.2 billion, with more to come over the course of the seventy-five-year contract. Without being given time to study the arrangement very deeply, the city council approved the deal. Six months later a report revealed that, according to an estimate made by the city's inspector general, had Chicago retained control of the meters and operated them itself, it could have realized almost $1 billion more than it would get over the course of the lease.

Under the agreement, Chicago had to pay the private partner for lost revenue whenever drivers were not allowed to park. This would occur, for example, when a street was closed for a parade or festival, when new bus or bicycle lanes necessitated the removal of parking meters, when street maintenance was being performed, or when a competing parking garage was built. Through 2014, the city was billed $61 million for such actions. This income was in addition to increased revenue from the meters that stayed in use, because the private partner was able to charge more per hour over longer hours of operation, something the city council would not likely have done because of concerns over citizen outrage and voter retribution.

Pittsburgh had a similar opportunity to privatize parking, but it was rejected by the city council. The experience of Chicago also played a role in Cincinnati, New York, and other cities turning down a trade of parking spaces for quick money. Chicago itself seemed to have learned from its mistakes by abandoning the idea of privatizing Chicago Midway International Airport, which is located in the heart of the city. States also have turned down tempting proposals. Pennsylvania rejected opportunities to receive billion-dollar offers to run its turnpike for seventy-five years (for $12.8 billion up front) and the state lottery for twenty years (for $34 billion). Those are surely tempting numbers to a state legislature struggling to balance a budget, but sometimes longer-term views do prevail.

Regardless of term, design is everything in a public-private partnership, from the crafting of the agreement to the operational decisions about how to manage the infrastructure that it governs. In 2006, for $3.8 billion on signing, the state of Indiana gave up for seventy-five years control of the 157-mile-long Indiana Toll Road to an international joint venture. A combination of increased tolls (by 76 percent over seven years) and decreased traffic (by 13 percent over the same period, no doubt in part due to the higher tolls) might have promised to keep the deal profitable for the Spanish and Australian investors, but Indiana state and local governments were in danger of reaping disastrous consequences. Tractor-trailer drivers who did not wish to pay exorbitant tolls began to take advantage of the many parallel and perpendicular roads that crisscross the flatness of rectilinearly divided northern Indiana.

This was not a new phenomenon, and there was even a term for it. An alternate route that travelers use to avoid the expense of using a toll road is known as a "shunpike," short for "shunning a turnpike." In the United States, shunpikes date from the early nineteenth century, and to this day there is a Shunpike Road in Chatham Township, New Jersey, which in 1804 paralleled the Morris Turnpike. Unfortunately, shunpike roads in northern Indiana were typically not designed to carry heavy truck traffic, which congested local routes, presented safety concerns, and generally affected the quality of life along the way. Furthermore, since the secondary roads could be expected to wear out faster under the heavier traffic, there would be increased maintenance costs. This was not likely anticipated by the representatives of the state when negotiating the P3 agreement, nor was the effect of canceling tolls to accommodate traffic diverted from a flooded nonprivatized highway. The state (read: its taxpayers) had to pay over $500,000 to the foreign operators as compensation for lost toll revenue. In spite of all the seeming advantages, in 2014 the toll road operator filed for bankruptcy. Before long, potential new investors were lining up to enter into a contract more favorable to them.

Pennsylvania provides another instructive example. On the infrastructure report card issued in 2014 by the Pennsylvania section of the American Society of Civil Engineers, the Keystone State received poorer grades for roads (D−) and bridges (D+) than the national averages (D and C+, respectively), and it had more structurally deficient bridges than any other state. To remedy the situation, the Pennsylvania Department of Transportation (PennDOT) considered the offers of four consortiums to finance, design, build, and maintain more than 500 of the state's 4,200 bridges said to be in need of repair or replacement. Although the state could do this kind of work itself, it would have been limited in how much money it could spend annually. By entering into a wide-ranging contract, the winning bidder was expected to produce generic designs and order common structural components in quantity, something the state could not do within its annual budget. Since the contract would call for penalties for inferior work, Pennsylvania expected to get long-lasting bridges faster and at a lower cost than it could otherwise.

The winning bid of $899 million for the "rapid bridge replacement project," came from the consortium of Plenary Walsh Keystone Partners, which agreed to replace 558 bridges and maintain them for twenty-five years. The work was scheduled to begin in the summer of 2015 and be completed within thirty-six months. Had PennDOT followed standard procedures and done all the work itself, it would have taken eight to twelve years to complete, and the average cost per bridge would have been more than $2 million as opposed to $1.6 million under the contract.

Pennsylvania's bridge replacement program was modeled after a Missouri initiative that repaired or replaced some eight hundred bridges in three and a half years, at a cost of $685 million. One of the features that made the project succeed was allowing the contractor to close a road and detour traffic rather than realign it where work was being done on a bridge. This not only saved the contractor the time and money it takes to reconfigure traffic lanes but also provided a better environment for the construction workers, who could do their jobs more safely and quickly. Had the state transportation department been doing the work, voter and other political pressures would surely have argued against closing a road to work on a bridge.

Critics of the various forms of public-private partnership have many objections to their being used by state and local governments. One claim is that the P3 agreement can "circumvent democracy" by "giving away urban planning ability" for streets and highways, the future use of which is not known for decades. If the private investor is allowed to make decisions regarding land use, for example, the public will not be able to exercise its right to vote on matters that can affect the entire community. As for which side of the partnership assumes the greater risk, this again depends upon the agreement made.

Another objection to P3 arrangements is that the private parties effectively receive government subsidies. These can take the form of federally guaranteed loans, which are backed by the U.S. Department of Transportation and allow deferment of payment for up to five years beyond project completion. This and various forms of tax-exempt financing that Congress has authorized provide a subsidy unique to the United States, which may explain the interest of foreign and domestic investors alike. According to a public interest research consultant, P3 investors belong to "the most corporate welfared-up industry that there is." Since the federally backed financing programs are relatively new, there is little experience with them, especially for projects with durations measured in a half century or more, and questions have been raised about the protection of taxpayers. Already, bankruptcies and defaults have occurred involving toll roads in California and in Texas as well as in Indiana.

Among the features that make public-private partnerships attractive to domestic departments of transportation is that they allow politicians to approve improvement projects without immediately raising taxes, tolls, or public debt. This is not necessarily a bad motive to choose a P3 alternative to build a new road or a new bridge. However, if the agreement does not protect the taxpayer from hidden costs—either because the project may become more expensive than if the state or local government had done it itself or in granting such liberal options for the private partner to garner ad hoc revenue—then perhaps the deal should not be made at all.

Another way to involve private investment in public works has been through the federal Immigrant Investor Program (also known as EB-5). Here, a foreign investor willing to risk money on an American infrastructure project can obtain permanent-resident status for himself and his family. The scheme was selected by the Pennsylvania Turnpike Commission to help fund a long-called-for direct route between the turnpike and Interstate 95, effectively connecting in a smooth fashion the Pennsylvania and New Jersey pikes. Chinese investors were targeted in this case, but others were welcome to participate. If four hundred individuals invested $500,000 each in the project, the $200 million raised would fund about half the estimated $420 million total cost. The balance would be paid for by the federal government and the turnpike commission. The investors would get their green cards, but they would not be guaranteed a positive return on their investment; what they would realize financially would depend on fluctuations in interest rates and the bond market. It is not unlikely that when money gets increasingly tight, federal and state governments and their agencies will come up with even more creative funding schemes for infrastructure projects.

Whether a public-private partnership is appropriate for a given infrastructure project will naturally depend on the circumstances and the details. If a government entity badly needs a new bridge, say, but also badly needs cash to balance its budget, a potential partnership with private investors can appear to be a godsend. However, if the investors or their negotiators see the government entity as being desperate to make a deal at any cost, then it could be taken advantage of. Ideally, public-private arrangements should be entered into by municipalities and states only when they are free to negotiate from a position of strength, meaning that they do not face an imminent fiscal or infrastructural crisis. The way to avoid being found in such a position is, of course, to plan well ahead so that partnerships are sought with plenty of time to negotiate a deal that is fair to both potential partners. Such an arrangement can indeed present a win-win situation.

21

Ages and Ages Hence

SMART CARS AND POTHOLE-FREE ROADS

THE FUTURE PROMISES TO bring us smart cars riding along smart high-ways and over smart bridges. Even if the drivers are not smart and nod off at the wheel, the futuristic vehicles they will be driving will find their way home the way a family horse did in olden days. As Robert Frost's little horse of yesteryear thought it queer to be stopping by woods on a snowy evening, wondering if there were some mistake and giving its harness bells a shake in inquiry, so the automobile of tomorrow prom-ises to question and warn us and even take control of matters when we try to do something foolish, like approach at full speed a car stopped ahead; move into a passing lane when another vehicle is on our flank but not visible in the side-view mirror; drift across the lane or edge mark-ings on the pavement; or drive past our exit.

Such driving aids are already available today, and the vehicle equipped with them communicates with the driver through a combination of aural, visual, and haptic prompts. Drivers become conditioned to and react to different numbers and tones of beeps, blinking and flashing lights and icons, or vibrating steering wheels and seats after a surpris-ingly short period of acclimation. Some of the systems are so sophisti-cated that they act when the driver does not, steering the car back into its lane or stopping the car automatically in an emergency. Some cars can park themselves, and a four-wheel-drive Tesla equipped with an autopilot will perform automatic lane changing when the driver signals his intention to do so. And, of course, in just about any car now, when we miss a turn the GPS navigator we are using will inform us that it is recalculating our route or that we should make a legal U-turn as soon as conditions allow.

Vehicles also are equipped with devices that can check the pressure in all four tires simultaneously and display on the now-ubiquitous center screen the results along with the correct inflation pressures. Sophisticated cruise control systems can not only maintain a steady speed but also can maintain a fixed number of safe car lengths between our vehicle and the one in front of us. This kind of smart control is achieved with such smoothness of deceleration that the driver having set the cruising speed at 65 miles per hour will realize that his speed has been reduced to 55 only when every other car on the highway is passing both him and the slowpoke in front of him.

Even on the darkest evening of the year, the headlights of today's cars brilliantly illuminate the dark and deep woods beside the road, some dimming themselves as traffic coming in the opposite direction warrants. When we wish to pull off the highway and onto a dirt road leading up to a farmhouse, the headlight beams will shift from pointing straight ahead to pointing left or right, depending on which way we are turning. Smart windshield wipers know to start working when a few drops of water strike the glass, and they adjust the speed of their sweep according to how heavily it is raining.

Experiencing such features in the cars of today, we might wonder what the future will bring. Will windshield wipers detect and clean off tree sap that streaks the glass? Will underinflated tires fill themselves even as we travel down the interstate? Will the illumination of the low-fuel warning icon trigger the GPS to identify the nearest service station selling our preferred brand of gasoline at the lowest local price and automatically direct us there? And if the distance to the pump is greater than the esti- mated miles of fuel left in the tank, will the car be slowed to the appropri- ate fuel-efficient speed to get us there before we run out of gas? These are the kinds of questions inventors and engineers pose to themselves on the way to improving present-day technology to make it future technology.

But the features just described have to do more with vehicles than with the roads and bridges upon which they rely. Already it is commonplace to have detectors embedded in the pavement at intersections to tell traffic lights whether there is a vehicle waiting to make a left turn. If there is not, the left-turn arrow does not appear and traffic coming the other way is given the green light without delay. There are also smart open roads and bridges, pieces of smart infrastructure that in essence monitor themselves and alert engineers and others when they are in need of attention. A roadway can be embedded with sensors that detect snow and ice conditions, signaling when road crews should be dispatched. Bridges can be fitted not only with sensors but also with devices controlled by the sensors. Thus, when ice begins to develop on a bridge surface, the sensors that detect it can also trigger a system embedded in the curb or guardrail that sprays anti-icing solution over the pavement without human intervention.

Another use of such technology is to fit a bridge with devices that can detect when a beam or girder in the structure has developed a serious crack or corrosion. It is expensive to wire a large number of such devices to a data-collection computer that can communicate with, say, the state department of transportation, but if the devices are wireless, the installation can be done much more easily and efficiently. One heavily used piece of infrastructure that would have benefited from detection devices is the Interstate 495 bridge over the Christina River in Wilmington, Delaware. In mid-2014, human inspectors found that a number of the piers supporting the approach viaduct were out of plumb and that the structure was leaning. The section of highway was closed while the problem was evaluated and, eventually, corrected. The viaduct, which ran over largely vacant industrial land, had been damaged when a 50,000-ton pile of dirt was illegally placed atop an area beside the bridge piers, causing the ground to shift and so damage the piers. Had they been fitted with sensitive wireless tilt meters akin to the kind that tell a smartphone or tablet which way is up, data indicating the abnormal condition might have been collected by a computer- and sensor-filled truck on a routine pass over the bridge. Earlier detection of the problem would likely have arrested the progressive damage and gotten the bridge back in service more quickly and at less expense.

In another case, in 2013, a bridge carrying Interstate 5 over the Skagit River in Washington State collapsed and fell into the water after an oversize truck hit parts of the overhead steel structure while passing through one of its truss spans. The fifty-eight-year-old bridge had been considered functionally obsolete, forcing traffic in the right-hand lane to drive very close to the side of the structure, where the overhead braces curved downward and so reduced vertical clearance. There had been a history of tall vehicles striking these structural parts, but the bridge had survived prior impacts. In addition to this problem, the Skagit River Bridge was also classified as fracture-critical, making it like the interstate bridge in Minneapolis that had collapsed six years earlier, claiming thirteen lives. No one died in the Skagit River Bridge collapse, but that was only a matter of luck.

Outfitting a bridge like the one over the Skagit River with sensors that detect a too-high oncoming vehicle would be easy to do. We already are familiar with devices that measure a vehicle's speed by radar and display it on a sign beside the highway, thereby providing a warning to the driver to slow down. The augmented cruise control systems that keep a car a safe distance from the one in front of it use lasers and a computer to maintain a specified separation. Imagine a smart bridge with low clearance fitted with a similar device that could also communicate with an approaching tall but smart truck and slow it down to a stop before it impacted the bridge. That kind of wireless connection between bridge structure and vehicle could obviously prevent collapses and deaths. In time, it is likely to be a common reality.

Engineers are currently working on highway-vehicle systems that maintain a constant wireless connection among vehicles within a few hundred yards of each other traveling along the same stretch of highway. In such a system, two cars not visible to each other—either because there is congested traffic between them or the one ahead is rounding a blind curve—would be able to maintain contact. If the leading car has to stop suddenly, it would send a signal to the following car to apply its brakes also—and do it automatically.

Other so-called connected-vehicle technology could lead to the elimination of stop, yield, and other traffic signs. If all vehicles in the vicinity of an upcoming intersection were part of the wireless network, the computer in a car approaching the intersection could know whether there would be cross traffic. If so, the driver would be alerted to approach with caution or stop, and perhaps even have a bright electronic stop sign displayed on a navigation screen, if not as a hologram directly in the driver's line of sight. If the intersection is anticipated to be clear, no warning would appear and the car would not even have to slow down. If all vehicles were part of such a system, traffic signs and lights could be eliminated entirely, thereby unburdening a transportation department of the need to purchase, install, and maintain such things. Furthermore, a highway without physical signs would be a less visually cluttered and hence a less distracting and more attractive and environmentally friendly road.

In the summer of 2014 the U.S. Department of Transportation announced its plan to require in the not-too-distant future the installation of vehicle-to-vehicle communication technology in all cars and trucks, new and old. Fitting a vehicle with a transmitter is projected to add about $350 to the cost of a new car in 2020; an existing car could be retrofitted with an equivalent device. Engineering research teams connected with universities, government laboratories, and the automotive industry are already testing the technology. As part of a program at the Michigan Transportation Research Institute in Ann Arbor, volunteers are driving almost three thousand transmitter-equipped vehicles in actual traffic conditions.

Completely driverless cars are already a reality, though so far only as prototype vehicles. The development of such vehicles has been promoted for years by the U.S. Department of Defense, which through its Defense Advanced Research Projects Agency has for some years been sponsoring driverless-car challenges, offering prizes as high as $2 million for unmanned vehicles successfully negotiating urban and battlefield courses. Building on the successes of such programs, Google has been working on the development of driverless automobiles since 2009, and in the subsequent five years such vehicles drove themselves autonomously for more than seven hundred thousand miles on public roads. The automotive industry has also been actively developing its own autonomous technology. An Audi Q5 outfitted with a computer system processing data from onboard instruments including cameras, radar,

and laser sensors in early 2015 completed a 3,400-mile trip from San
Francisco to New York City, during which it behaved as an autonomous
vehicle and steered itself 99 percent of the time. It was only in very
complicated traffic situations like construction zones that a human
driver had to take over the wheel. Tesla Motors had plans to offer its
Model S electric-powered sedan in a self-steering version in the summer
of 2015, and Elon Musk, chief executive officer of Tesla, promised a
fully autonomous vehicle by about 2020. General Motors is expected
to introduce in 2017 technology in its Cadillac that will allow no-hands
highway driving. The configuration of cars will naturally evolve with
the use of such technology, and in time we can expect front seats
to swivel around so that, if they wish, everyone in the vehicle can play
card and board games, converse face-to-face, or catch up on e-mail,
send text messages, and otherwise immerse themselves in antisocial
media.

Even when a vehicle's occupants do not have their eyes on the road,
they can help inspect and maintain it by being "data donors." This is a
concept introduced by New Urban Mechanics, a group dedicated to "an
approach to civic innovation focused on delivering transformative city
services to residents." In 2010, the Mayor's Office of New Urban
Mechanics was formed in Boston, later joined by an analogous city
agency in Philadelphia, each serving as an "innovation incubator" by
"building partnerships between internal agencies and outside entrepreneurs to pilot projects that address resident needs." Specific projects
ranged from improved trash cans to high-tech smartphone apps.

Among the apps is SmartBump, which tracks an automobile's ride
over Boston streets. After the free app has been activated, the smartphone can be placed on a car's dashboard to monitor the ride. The
accelerometer in the phone detects bumps, and software distinguishes a
bump due to hitting a pothole from one due to riding over a manhole or
a speed bump. Where the vehicle does go over a pothole, the detected
movement is paired via GPS with the location, and the data are sent—
this is the data donation—to the appropriate city department that
monitors the condition of the streets. When a pothole is identified, a
road crew is sent to fill it. Engineers at Northeastern University have

taken the concept a bit further with their Versatile Onboard Traffic Embedded Roaming Sensors, a mouthful no doubt forced to yield the acronym VOTERS. Vehicles driving a lot of city streets each day are outfitted with a variety of sensors that measure such things as tire sound, pavement surface defects, and subsurface delamination—data that when properly interpreted provide information on precursors to potholes. Suspect conditions can be monitored and repairs done when there are signs of a pothole beginning to form.

Potholes themselves may be a thing of the past if researchers succeed in developing self-healing asphalt. One of the most prominent of these researchers is Erik Schlangen, a professor of civil engineering at Delft University of Technology in the Netherlands and leader of a group doing research in experimental micromechanics. By adding short steel-wool fibers into a mix of asphalt concrete, cracks that develop in the pavement material can be healed when it is subjected to microwaves. Schlangen has demonstrated the process in a six-and-one-half-minute TED talk, in which he breaks a small beam made of asphalt in two, places the two halves together in a microwave oven, and by the end of his talk removes a rejoined bar. His hope is to develop an industrial-scale microwave device that can be transported in a highway vehicle and perform the same kind of healing by induction.

Self-healing concrete pavement made from Portland cement is also on the horizon. It can be made by incorporating into the concrete mix tiny capsules containing dormant bacteria capable of producing limestone when activated. This will happen when the concrete develops a crack that causes the encapsulated bacteria to be released and that allows water to reach the bacteria, thus enlivening them to do their work. Once activated, the bacteria consume a starchy substance that was incorporated into the concrete mix and excrete calcium carbonate, which is essentially limestone, thus plugging the crack and forestalling further damage. Structures formed of such concrete have recovered as much as 90 percent of their uncracked strength.

The roads of the future promise to be smooth and quiet, as will be the ride in all-electric vehicles that occupants will no doubt take for granted. The experience, however, may not be quite the same as Norman

Bel Geddes laid out in the General Motors Futurama exhibit and in his book, *Magic Motorways*. It is likely to be far superior.

WHETHER OUR ROADS AND bridges can actually evolve to such an idyllic state in the foreseeable future will depend upon how well we care for them in the interim. If America's highway infrastructure is allowed to deteriorate much below its current state, the cost of just maintaining it in a condition no worse than it is now could be overwhelming. Almost every dollar budgeted for roads and bridges, whether at the federal, state, or local level, will have to be earmarked for repair and replacement work just to restore the status quo ante rather than advance the technology and apply it broadly. Government-sponsored research and development of smart materials and systems will have to take a backseat to filling potholes, resurfacing roads, and rebuilding broken bridges. And the longer we wait to do those things, the more there will be to fill, resurface, and rebuild, which will of course take even more money.

Since first issued in the late 1990s, the infrastructure report cards now updated every four years by the American Society of Civil Engineers have publicized the dire state of affairs. Ironically, the category of bridges, which have become symbols for the assortment of public works collectively termed "infrastructure," has received among the highest grades. But a gentleman's C, defined by ASCE itself as "mediocre," should not be the nation's goal. It should signal plenty of room for improvement, not only in aspirations toward better-quality materials and workmanship but also toward more durable structures and systems. Neither voters nor elected officials should be satisfied with or tolerate the mediocre to poor infrastructure that we are told our nation now possesses. In fact, we should be embarrassed. We as a nation should want our public works to be better than they are.

The country has pulled itself out of infrastructural ditches before. The Good Roads Movement begun by cyclists in the late nineteenth century led to improved roads that were enjoyed by recreational cyclists and early motorists alike. The firsthand experience of Lieutenant

Colonel Dwight Eisenhower in an Army convoy and of other pioneers who embarked on transcontinental journeys in early motor vehicles brought to the fore the need for national roads. In the 1920s, the plight of farmers struggling to get their wares to market over unimproved rural roads attracted the attention of champions in Washington, who pulled the farmers out of the mud and onto an improved system of farm-to-market roads. And in the 1930s, even as federal road-building projects were providing jobs during the Depression, plans were advancing for a network of interregional highways that in the latter 1950s began to be realized in the interstate highway system. But by 2006, the year the fiftieth anniversary of the landmark interstate legislation was celebrated, there was a general recognition that the interstates themselves were growing old and in some places had become inadequate to their task. The American Recovery and Reinvestment Act of 2009, touted as a stimulus package for the economy, was very visibly credited on large signs announcing highway improvement projects it funded, but the total amount of money devoted to them was barely 4 percent of the bill's approximately $800 billion total. We are still in an infrastructural ditch.

According to the American Society of Civil Engineers report card for 2013, the average age of America's bridges was forty-two years, and more than 30 percent of the spans were beyond their fifty-year design life. In times of tight budgets, there is necessarily a tension between the increasing cost of maintaining aging bridges in service and replacing perhaps even younger ones that are structurally deficient. Collectively, the latter group includes about 10 percent of all bridges in the United States, and these carry approximately a quarter of a billion vehicles daily. Occasionally such a bridge will collapse without warning, as the one carrying Interstate 35W across the Mississippi River in Minneapolis did during an evening rush hour in 2007. It is to preclude such tragic accidents that conscientious surveillance, maintenance, and replacement of bridges and other existing infrastructure is obviously so important.

Properly maintaining and, when and where needed, upgrading and expanding our infrastructure not only ensures that we have safe roads and bridges but also contributes positively to our quality of life, to our outlook for the future, and to the national economy. Shortchanging our investment in the infrastructure is shortchanging the well-being and optimism of future generations and the prospect for economic growth. Especially during slow economic times, allocating resources for infrastructure projects can provide not only much-needed jobs for the unemployed but also conspicuous productive activity at construction sites, thus promoting an ethic of work and progress even in the leanest of times.

We citizens can do our part to promote safe and robust infrastructure by actively and enthusiastically encouraging and supporting legislation that provides appropriate funding for infrastructure needs, replenished as needed through adequate and reliable sources of revenue. These generally will take the form of taxes and fees, which when fairly levied should be responsibly paid. For roads and bridges, as in all infrastructure categories, we should look to making the means of funding stable and predictable, so that state highway departments need not engage in triage that forces them to choose between whether to replace a structurally deficient bridge that is beyond its design life or to build a new and wider bridge to carry more traffic through a congested area.

No one likes to pay taxes, but they obviously are necessary to give us the roads and bridges and other forms of infrastructure that we have come to expect and appreciate. But expecting to drive over smooth pavements and adequately wide bridges should not be seen as a luxury that we can casually forgo in tough times. Poor roads and bridges actually cost households and businesses money in terms of increased commuting time, addition fuel use, and larger vehicle maintenance bills, in addition to higher prices for consumer goods whose distribution costs are higher. According to the ASCE *Failure to Act* estimate, the nation's degrading infrastructure could cost American households over the period from 2012 to 2040 in excess of $150 trillion, not accounting for inflation. That alone should be reason enough to support investment in infrastructure.

I was certainly not aware of such dilemmas when I was a young child playing on the edge of Brooklyn infrastructure in the aftermath of a thunderstorm. Squatting on the curb between sidewalk and street, my only focus was on the storm water flowing swiftly in the gutter. There was no need for me to think of the broader implications of where the water went after it passed by me and spilled into the sewer. Nor was there any need for me to know anything about the granite curbstone that separated the concrete sidewalk and asphalt pavement. As long as it supported me in my play, I was happy. It did not matter to me then that the street on which I played ran down to the Gowanus Canal, about a mile away. As children we measured distance in city blocks, and a mile was too far for a four-year-old urbanite to contemplate. Since the Brooklyn Bridge was about two miles away from my house, it might as well have been on a different planet. It did not capture my imagination then that older children and adults on foot or wheels could navigate along the streets and avenues and through the subway tunnels in the vicinity of my house all the way through downtown Brooklyn and on into Lower Manhattan, even though I knew that my father did it every workday.

By the time I was a teenager and drove my father's car past the construction sites of the Verrazano-Narrows and Throgs Neck bridges, I did begin to understand the interrelatedness of infrastructure. I understood that these bridges could not be built without consideration of the roads that brought traffic to them and carried it away from them. Infrastructure was like skeletal anatomy: the street is connected to the avenue; the avenue to the boulevard; the boulevard to the bridge; the bridge to the highway; and the highway to the world beyond. I also began to understand that these pieces of infrastructure did not last forever. Even the granite curb would wear down with use.

Growing up and stepping back from the curb provides a broader perspective. Reflecting on their history, it is easier to understand where roads come from and where they lead to, how they got their lines both white and yellow, how traffic on them was tamed with signs and lights, and why today's roads so often trace the same path as bison did in early America. The story of infrastructure, even in its smallest details, is one of growth and change in response to problems, but it is also one of expansion of systems and repetition of process. The problems of the

past remain the problems of today, only embedded in different technology, complexity, and bureaucracy.

Just as a poem written a century ago about making choices can
remain relevant today, so do the stories of the design, planning, and
construction of roads and bridges from earlier times. Whatever roads
have been taken or not taken to get us to the present state of our infrastructure, the lessons of the past can help us better comprehend how we
can put it back in order.

Acknowledgments

In addition to the references specific to each chapter given below, among the resources most valuable to me in writing this book were the daily online compilations of topical news distributed through the American Society of Civil Engineers' SmartBrief, *Engineering News-Record*'s Enr.com, and the American Association of State Highway and Transportation Officials' Daily Transportation Update and weekly Journal. Where I have expanded on what I had previously written and published in my engineering column in *American Scientist* or elsewhere, I have listed the pertinent bibliographic information among other references for the relevant chapter.

As usual, I have benefited greatly from the resources and services of the Duke University Libraries, including Document Delivery and the institution of Interlibrary Loan, and their respective staffs. These conveniences, coupled with the seemingly endless benefits of the Internet and World Wide Web, including digitally scanned versions of obscure historical documents, have made doing research on matters of historical and current interest practically an armchair activity. I am grateful to the often anonymous staff members who made such things possible.

When I proposed the idea for this book to George Gibson, he expressed an immediate interest in it, for which I am grateful. His editing of my submitted manuscript was perceptive, wise, and sensitive, no doubt saving the reader from having to endure many digressions. It has been a pleasure working with George, with senior production editor Sara Kitchen, and with their colleagues at Bloomsbury.

Catherine Petroski has played an indispensable role in the realization of this book. In addition to being my first reader, she has offered extremely helpful suggestions at every stage from rough draft to final manuscript. She also performed magic in helping me secure images and turning those that needed work into files of publishable quality. As always, thank you, Catherine.

Bibliography

1. THE ROAD TAKEN

Badger, Emily. "Joe Biden Is Very Angry at Us for Scrimping on Our Infrastructure," *Washington Post*, Wonkblog, October 21, 2014, http://www.washingtonpost.com/blogs/wonkblog/wp/2014/10/21/joe-biden-is-very-angry-at-us-for-skrimping-on-our-infrastructure/.

Frost, Robert. *Mountain Interval* (New York: Henry Holt, 1916), p. 9. See also http://www.poetryfoundation.org/poem/173536.

2. THE ROAD NOT TAKEN

American Society of Civil Engineers. *2013 Report Card for America's Infrastructure* (Reston, Va.: ASCE, 2013), http://www.asce.org/Infrastructure/Report-Card/Report-Card-for-America%E2%80%99s-Infrastructure/.

Baker, Michael. *Rebuilding America's Infrastructure: An Agenda for the 1980s* (Durham, N.C.: Duke University Press, 1984).

Choate, Pat, and Susan Walter. *America in Ruins: Beyond the Public Works Pork Barrel* (Washington, D.C.: Council of State Planning Agencies, 1981).

———. *American in Ruins: The Decaying Infrastructure* (Durham, N.C.: Duke University Press, 1983).

Hartgen, David T., M. Gregory Fields, and Baruch Feigenbaum. *21st Annual Report on the Performance of State Highway Systems (1984–2012)*, Policy Study 436 (Los Angeles: Reason Foundation, 2014), http://www.hartgengroup.net/Projects/National/USA/21st_annual/2014-09-12-21st_Annual_Highway_Report.pdf.

Hartgen, David T., M. Gregory Fields, and Elizabeth San José. "Examining 20 Years of U.S. Highway and Bridge Performance Trends," Reason Foundation, February 21, 2013, http://reason.org/news/show/1013203.html.

Melman, Seymour. "Looting the Means of Production," *New York Times*, July 26, 1981, op-ed page, http://www.nytimes.com/1981/07/26/opinion /looting-the-means-of-production.html.

National Council on Public Works Improvement. *Fragile Foundations: A Report on America's Public Works* (Washington, D.C.: U.S. Government Printing Office, 1988).

Petroski, Henry. "Infrastructure," *American Scientist*, September-October 2009, 370-374.

Strunsky, Steve. "Bayonne Bridge Races Against Arrival of Super-sized Ships," NJ.com, November 17, 2014, http://www.nj.com/news/index.ssf/2014/11 /officails_say_bayonne_bridge_roadway_raising_is_on_schedule_25 _percent_complete.html.

3. ROADS

Brown, Jeff L. "Rocky Road: The Story of Asphalt Pavement," *Civil Engineering*, May 2013, 40–43.

Christie, Deborah Carnes. *Green House: The Story of a Healthy, Energy-Efficient Home* (Durham, N.C.: privately printed, 2009).

Holley, I. B., Jr. *The Highway Revolution, 1895–1925: How the United States Got Out of the Mud* (Durham, N.C.: Carolina Academic Press, 2008).

Judson, William Pierson. *City Roads and Pavements Suited to Cities of Moderate Size*, second edition and fourth edition, revised (New York: Engineering News Publishing Co., 1902 and 1909).

Lay, Maxwell G. *Ways of the World: A History of the World's Roads and of the Vehicles that Used Them* (New Brunswick, N.J.: Rutgers University Press, 1992).

Longfellow, Rickie. "Back in Time: the National Road," Highway History, Federal Highway Administration, http://www.fhwa.dot.gov/infrastructure /back0103.cfm.

McShane, Clay. "Transforming the Use of Urban Space: A Look at the Revolution in Street Pavements, 1880–1924," *Journal of Urban History* 5, no. 3 (May 1979): 279–307.

Murphy, William. "Better Asphalt No Shortcut to Pothole-Free Roads," (New York) *Newsday*, March 9, 2014.

Petroski, Henry. "Twin Bridges," *American Scientist,* January–February 2001, 15–19.

Ransome, Ernest Leslie. "Concrete Construction," U.S. Patent No. 516,113 (March 6, 1894).

Ransome Construction Co. *Ransome Patent Concrete-Cold-Twisted-Steel Construction* (Philadelphia: privately printed, c. 1910).

Sosoff, Howard. "Transportation Infrastructure is the Road to More Competitive US Manufacturing," *Industry Week*, November 5, 2014, http://www.industryweek.com/transportation/transportation-infrastructure-road-more-competitive-us-manufacturing.

Weightman, Gavin. *Eureka: How Invention Happens* (New Haven, Conn.: Yale University Press, 2015).

4. DIVERGED

American Association of State Highway and Transportation Officials. "Interstate 50th Anniversary," 2006, http://www.interstate50th.org.

———. *The States and the Interstates: Research on the Planning, Design and Construction of the Interstate and Defense Highway System* (Washington, D.C.: AASHTO, 1991).

[Becker, Chris.] *AASHTO 1914–2014: A Century of Achievement for a Better Tomorrow* ([Washington, D.C.]: AASHTO, 2014).

Davies, Pete. *American Road: The Story of an Epic Transcontinental Journey at the Dawn of the Motor Age* (New York: Henry Holt, 2002).

Hayes, Brian. *Infrastructure: A Field Guild to the Industrial Landscape* (New York: Norton, 2005).

Lewis, Tom. *Divided Highways: Building the Interstate Highways, Transforming American Life* (New York: Viking, 1997).

Lin, James. "Lincoln Highway," 1998, http://www.ugcs.caltech.edu/~jlin/lincoln/history/part4.html.

Patton, Phil. "Road Signs of the Times," *New York Times*, January 21, 2005, D8.

Reid, Robert L., Laurie A. Shuster, and Jay Landers. "Special Report: The Interstate Highway System at 50," *Civil Engineering*, June 2006, 36–39.

Swift, Earl. *The Big Roads: The Untold Story of the Engineers, Visionaries, and Trailblazers Who Created the American Superhighways* (Boston: Houghton Mifflin Harcourt, 2011).

Weingroff, Richard F. "The Origins of the U.S. Numbered Highway System," *AASHTO Quarterly*, Spring 1997, 6–15.

White, L. A. "Dangers of the Unmarked Road," *Highway Magazine*, August 1919, 7–9.

5. YELLOW

[Becker, Chris.] *AASHTO 1914–2014: A Century of Achievement for a Better Tomorrow* ([Washington, D.C.]: AASHTO, 2014).

Curtiss, Aaron. "Shedding Light on History of Traffic Signals," *Los Angeles Times*, April 3, 1995. http://articles.latimes.com/1995-04-03/local/me -50360_1_traffic-light.

Eno, William Phelps. *Street Traffic Regulation* (New York: Rider and Driver Publishing, 1909).

———. *The Story of Highway Traffic Control: 1899–1939* ([Westport, Conn.]: Eno Foundation for Highway Traffic Control, 1939), http://ntl.bts .gov.

Federal Highway Administration. "The Evolution of MUTCD," *Manual on Uniform Traffic Control Devices*, http://mutcd.fhwa.dot.gov.

Gray, Christopher. *Fifth Avenue, 1911, from Start to Finish* (New York: Dover, 1994).

———. "The Beautiful Dawn of Traffic Signals," *New York Times*, May 18, 2014, RE11.

———. "Mystery of 104 Bronze Statues of Mercury," *New York Times*, February 2, 1997, Real Estate page.

Greenbaum, Hilary, and Dana Rubinstein. "The Stop Sign Wasn't Always Red," *New York Times Magazine*, December 9, 2011, MM30.

Hawkins, H. Gene, Jr. "Evolution of the U.S. Pavement Marking System," prepared as part of an Interim Report for NCHRP Project 4-28 — Feasibility Study for an All-White Pavement Marking System (College Station, Texas: Texas Transportation Institute, 2000).

Hoge, James B. "Municipal Traffic-Control System," U.S. Patent No. 1,251,666 (January 1, 1918).

Howard, Ralph. "Fifth Avenue's Traffic Tower," *Scientific American*, January 15, 1921, 45, 57–58.

"John Van Nostrand Dorr Dies; Metallurgist, 90, Was Inventor," *New York Times*, July 1, 1962, 57.

Kaszynski, William. *The American Highway: The History and Culture of Roads in the United States* (Jefferson, N.C.: McFarland, 2000).

[Kulsea, Bill, and Tom Shawver.] *Making Michigan Move: A History of Michigan Highways and the Michigan Department of Transportation* ([Lansing]: Michigan Department of Transportation, [1992]).

McClintock, Miller. *Street Traffic Control* (New York: McGraw-Hill, 1925).

McShane, Clay. "The Origins and Globalization of Traffic Control Signals," *Journal of Urban History* 25, no. 3 (March 1999): 379–404.

Montgomery, John A. *Eno, the Man and the Foundation: A Chronicle of Transportation* (Westport, Conn.: Eno Foundation for Transportation, 1988).

Morgan, Garrett A. "Traffic Light," U.S. Patent No. 1,475,024 (November 20, 1923).

Rice, Diana. "Beautiful Bronze Towers for Fifth Avenue Traffic," *New York Times*, November 5, 1922, 110.

"Right Shoulder Guides Gain as Road Safeguards," *Fleet Owner*, July 1955, 128–29.

Ross, Greg. *Futility Closet: An Idler's Miscellany of Compendious Amusements* (Raleigh, N.C.: Futility Closet Books, 2014).

Sawyer, K. I. "Handling Motor Truck Traffic on Trunk Highways of Marquette County, Mich.," *Municipal and County Engineering* 59, no. 3 (September 1920): 94–95, http://books.google.com/books?id=i85LAAAAYAAJ&pg=PA94&lpg=PA94&dq=%22rapidly+becoming+as+serious+a+problem%22&source=bl&ots=VAUZxJLKFR&sig=Q6ACk2shLgNst.Yz-_BEAuBVrApM&hl=en&sa=X&ei=f3BiVJbtEYqsyASDoIHQDA&ved=0CCIQ6AEwAg#v=onepage&q=%22rapidly%20becoming%20as%20serious%20a%20problem%22&f=false.

Stuff Nobody Cares About. "Old New York in Photos #19: Traffic Signal—Fifth Avenue and 34th Street, 1922," http://stuffnobodycaresabout.com/2012/06/27/old-new-york-in-photos-19/.

Thiusen, Ismar. *The Diothas; or, A Far Look Ahead* (New York: G. P. Putnam's Sons, 1883).

"Traffic Expert Opposes Towers," *New York Times*, April 9, 1922, 44.

"Traffic-Marks on Country Roads," *Literary Digest*, October 23, 1920, 29.

6. COULD NOT TRAVEL

For having kept me posted, via e-mail messages and newspaper clippings, on the ongoing saga of the East Bay Crossing, I am grateful to many Bay Area residents—especially Bob Piper of Berkeley, Peter Neumann of Menlo Park, and Karen Petroski, then of San Francisco. I am grateful to Paul Giroux for arranging an invaluable inside-the-bridge tour of the new East Bay viaduct of the San Francisco–Oakland Bay Bridge while it was still under construction. Thanks to TJ Bross for his help in identifying images of the signature span of the new bridge, and to Maribel Castillo and Nina Chan of T.Y. Lin International for providing the photo showing the new bridge with the partially disassembled old cantilever in the background.

Bryant, Elizabeth. "A Tale of Two Bridges," *San Francisco Chronicle*, December 24, 2004, A1, A9.

California Department of Transportation. *San Francisco–Oakland Bay Bridge East Span Seismic Safety Project,* 2009, http://www.dot.ca.gov/dist4 /eastspans/index.html.

Camo, Sante. "The Evolution of a Design," *Structural Engineer*, January 2004, 32–37.

Dillon, Richard. *High Steel: Building the Bridges Across San Francisco Bay* (Millbrae, Calif.: Celestial Arts, 1979).

Dresden, Matthew. "Must a Bridge Be Beautiful Too?" *Access*, no. 28 (Spring 2006), 10–17.

Herel, Suzanne. "Emperor Norton's Name May Yet Span the Bay," *San Francisco Chronicle*, December 15, 2004, A1.

MacDonald, Donald, and Ira Nadel. *Bay Bridge: History and Design of a New Icon* (San Francisco: Chronicle Books, 2013).

Metropolitan Transportation Commission. *Planning: San Francisco–Oakland Bay Bridge,* 2009, http://www.mtc.ca.gov/planning /bay_bridge.

Mladjov, Roumen V. "The Most Expensive Bridge in the World," *Modern Steel Construction*, September 2004, 53–55.

Plummer, John W. *The World's Two Greatest Bridges* (San Francisco: Personal Stationery Company, 1936).

Pollak, Daniel. *Timeline of the San Francisco–Oakland Bay Bridge Seismic Retrofit: Milestones in Decision-Making, Financing, and Construction.* California Research Bureau Report No. CRB 04-013 (Sacramento, Calif.: California State Library, 2004), https://www.library.ca.gov/crb/04/13/04 -013.pdf.

Powers, Mary B. "California Scraps Sole Bid for Signature Span," *Engineering News-Record*, October 11, 2004, 10–11.

Princeton University. *Teaching and Scholarship in the Grand Tradition of Modern Engineering: A Symposium in Honor of David P. Billington at Seventy-Five and Forty-Five Years of Teaching at Princeton University* (Princeton, N.J.: Department of Civil and Environmental Engineering, 2004).

Romney, Lee. "Calif. Senate Criticizes Bay Bridge Construction Oversight in Two Reports," *Los Angeles Times*, August 1, 2014, via ENR.com, http

://enr.construction.com/yb/enr/article.aspx?story_id=id:XZ7KBm2ZVrT6o
PnkeBUCkwz2UxAJ3xd4g8A8YoergnBSXOWgRHX_o_Prml9AwEs9.

Schneider, Keith. "On Books Since 1988, Ohio River Dam Project Keeps Rolling Along," *New York Times*, August 19, 2014, A10, A15.

Van Derbeken, Jaxon. "Latest Defect: Bay Bridge Tower Rods Sitting in Water," SFGate, October 3, 2014, http://www.sfgate.com/bayarea/article/Nearly -all-Bay-Bridge-tower-rods-sitting-in-water-5792535.php.

———. "New Bay Bridge Plan: Turn Piers into Park," SFGate, November 28, 2014, http://www.sfgate.com/bayarea/article/New-Bay-Bridge-plan-Turn -piers-into-park-5923505.php.

———. "Fixing Bay Bridge Tower's Lean Put Crucial Rods at Risk," SFGate, March 6, 2015, http://www.sfgate.com/bayarea/article/Fixing-Bay-Bridge -tower-s-lean-put-curcial-rods-6118066.php.

Vorderbrueggen, Lisa. "Building the Bay Bridge: 1930s vs. Today," *San Jose Mercury News*, August 10, 2013, http://www.mercurynews.com/breaking -news/ci_23833904/building-bay-bridge-1930s-vs-today.

7. IN THE UNDERGROWTH

Bel Geddes, Norman. *Magic Motorways* (New York: Random House, 1940).

California Department of Transportation. "Caltrans Celebrates its 120th Anniversary: Bicyclists to Thank for Helping Make Highway System a Reality," press release, April 8, 2015, http://www.dot.ca.gov/hq/paffairs /news/pressrel/15pro34.htm.

Eno, William Phelps. *The Story of Highway Traffic Control: 1899-1939* ([Saugatuck, Conn.]: Eno Foundation for Highway Traffic Control, 1939), http://ntl.bts.gov.

General Motors. *To New Horizons*, 1940 film, https://archive.org/details /ToNewHor1940.

Hood, Clifton. *722 Miles: The Building of the Subways and How They Transformed New York* (Baltimore: Johns Hopkins University Press, 1993).

Johnson, Kirk. "Reanimating Bertha, a Mechanical Behemoth Slumbering Under Seattle," *New York Times*, August 2, 2014, A11.

Lindblom, Mike. "Is Bertha's Pit Affecting Pioneer Square Buildings?" *Seattle Times*, December 8, 2014, http://seattletimes.com/html/localnews /2025193787_viaductsettlingxml.html.

——. "Bertha Repair Will Take Longer—There Are Lots of Broken Parts," *Seattle Times*, May 19, 2015. http://www.seattletimes.com/seattle-news /transportation/bertha-repair-will-take-longer-theres-more-damage.

Malcolm, Tom. *William Barclay Parsons: A Renaissance Man of Old New York* (New York: Parsons Brinckerhoff, 2010).

——. "The Renaissance Man of New York's Subways," *TR News*, January– February 2006, 8–13.

McClintock, Morris. *Street Traffic Control* (New York: McGraw-Hill, 1925).

McShane, Clay. "The Origins and Globalization of Traffic Control Signals," *Journal of Urban History* 25, no. 3 (March 1999): 379–404.

Montgomery, John A. *Eno, the Man and the Foundation: A Chronicle of Transportation* (Westport, Conn.: Eno Foundation for Transportation, 1988).

Parsons, W. B., Jr. *Turnouts: Exact Formulae for their Determination, Together with Practical and Accurate Tables for Use in the Field.* (New York: Engineering News Publishing, 1884).

——. *Track, a Complete Manual of Maintenance of Way, According to the Latest and Best Practice on Leading American Railroads* (New York: Engineering News Publishing, 1886).

Parsons, William Barclay. *The American Engineers in France* (New York: Appleton, 1920).

——. *Engineers and Engineering in the Renaissance* (Baltimore: Williams & Wilkins, 1939).

Parsons, Wm. Barclay. *Report to the Board of Rapid Transit Railroad Commissioners in and for the City of New York on Rapid Transit in Foreign Cities* ([New York]: 1894).

——. *An American Engineer in China* (New York: McClure, Phillips, 1900).

——. *Robert Fulton and the Submarine* (New York: Columbia University Press, 1922).

Petroski, Henry. "William Barclay Parsons," *American Scientist*, July 2008, 280–83.

Suplee, Henry Harrison. "A Five-Storied Street," *Cassier's Engineering Monthly* 43, no. 6 (June 1913): 57–60.

Swift, Earl. *The Big Roads: The Untold Story of the Engineers, Visionaries, and Trailblazers Who Created the American Superhighways* (Boston: Houghton Mifflin Harcourt, 2011).

Walker, James Blaine. *Fifty Years of Rapid Transit, 1864 to 1917* (New York: Law Printing, 1918).

8. JUST AS FAIR

This chapter began as a column based on reports that appeared in a wide variety of newspapers and trade journals, including the *New York Times* and *Engineering News-Record*. I am grateful to Marianne Petroski for sending me clippings from the *Rockland County Journal News*, which has continued to cover the ongoing story of the bridge from a very interested local perspective. Michael LaViolette and Marcia Earle of HDR Engineering, as well as Carla Julian of Tappan Zee Constructors, pointed me to the source of images of the old and new Tappan Zee crossing made available through the New York State Thruway Authority.

LaViolette, Michael D. "Tappan Zee Hudson River Crossing," *Structure*, October 2014, 27–29. http://www.structuremag.org/?p=7098.

New York State. "The New NY Bridge," http://www.newnybridge.com/.

Petroski, Henry. "Tappan Zee Bridge," *American Scientist*, May–June 2012, 172–76.

Yee, Vivian. "Linked by a Bridge, and a Myth," *New York Times*, November 22, 2014, A16, A17.

9. PERHAPS THE BETTER CLAIM

Carter, Matt. "The Business of Beautiful Bridges," *ArupConnect*, November 10, 2014, http://www.arupconnect.com/2014/11/10/the-business-of-beautiful -bridges.

"Controversial Architect Santiago Calatrava Defends His Record," *Architectural Record*, June 9, 2015, http://archrecord.construction.com/yb/ar/article .aspx?story_id=id:jqRL65Amw2p875gVB4JSocitjhkBHnODKuDRRz MSLK3Un_n-qWHqvZ8gCR4JoYi8.

Csogi, Ralph D. "Reconstructing the Manhattan Bridge," *Civil Engineering*, January 2015, 56–65.

Dunlap, David W. "How a Train Station's Price Swelled to $4 Billion," *New York Times*, December 3, 2014, A1, A24–A25.

Dwyer, Jim. "A Stunning Link to New York's Past Makes a Long-Awaited Return," *New York Times,* June 5, 2015, A19.

Flyvbjerg, Bent. "What You Should Know About Megaprojects and Why: An Overview," *Project Management Journal*, April/May 2014, http://papers .ssrn.com/sol3/papers.cfm?abstract_id=2424835.

Flyvbjerg, Bent, Nils Bruzelius, and Werner Rothengatter. *Megaprojects and Risk: An Anatomy of Ambition* (Cambridge: Cambridge University Press, 2003).

Guastavino, Rafael. "Construction of Tiled Arches for Ceilings, Staircases, &c.," U.S. Patent No. 430,122 (June 17, 1890).

Hadlow, Robert W. *Elegant Arches, Soaring Spans: C. B. McCullough, Oregon's Master Bridge Builder* (Corvallis: Oregon State University Press, 2001).

Hayes, Brian. *Infrastructure: A Field Guide to the Industrial Landscape* (New York: W. W. Norton, 2005).

Leslie, Jacques. "The Trouble with Megaprojects," *New Yorker*, April 11, 2015, www.newyorker.com/news/news-desk/bertha-seattle-infrastructure-trouble -megaprojects.

Measuring Worth. "Seven Ways to Compute the Relative Value of a U.S. Dollar Amount—1774 to Present," http://www.measuringworth.com/uscompare/.

Petroski, Henry. "Bridging the Gap," *New York Times Magazine*, June 14, 2009, 11–12.

———. *Engineers of Dreams: Great Bridge Builders and the Spanning of America* (New York: Alfred A. Knopf, 1995).

———. "Tappan Zee Bridge," *American Scientist*, May–June 2012, 172–76.

Talese, Gay. *The Bridge* (New York: Bloomsbury, 2014).

10. BECAUSE IT WAS

American Association of State Highway and Transportation Officials. *Roadside Design Guide*, 4th edition (Washington, D.C.: AASHTO, 2011).

Ivory, Danielle, and Aaron M. Kessler. "Highway Guardrail May Be Deadly, States Say," *New York Times*, October 13, 2014, A1, B4.

———. "U.S. Accepts Trinity's Plan to Test Design of Guardrail," *New York Times*, November 12, 2014, B6, http://www.nytimes.com/2014/11/13 /business/us-accepts-plan-for-new-tests-of-trinity-guardrail-design.html.

Kessler, Aaron M., and Danielle Ivory. "Virginia to Remove Suspect Guardrails," *New York Times*, October 28, 2014, B1, B4.

Kozel, Scott M. "New Jersey Median Barrier History," *Roads to the Future*, 2004, http://www.roadstothefuture/Jersey_Barrier.html.

McDevitt, Charles F. "Basics of Concrete Barriers," *Public Roads*, 63, no. 5 (March/ April 2000), http://www.fhwa.dot.gov/publications/publicroads/oomarapr /concrete.cfm.

Ray, Malcolm H., and Richard G. McGinnis. *Synthesis of Highway Practice 244: Guardrail and Median Barrier Crashworthiness* (Washington, D.C.: National Academies Press, 1997).

Waldman, Jonathan. *Rust: The Longest War* (New York: Simon & Schuster, 2015).

11. GRASSY AND WANTED WEAR

American Association of State Highway and Transportation Officials. *The States and the Interstates: Research on the Planning, Design and Construction of the Interstate and Defense Highway System* (Washington, D.C.: AASHTO, 1991).

Bassett, Edward M. "The Freeway—A New Kind of Thoroughfare," *American City*, February 1930, 95.

[Bureau of Public Roads], *Toll Roads and Free Roads* (Washington, D.C.: U.S. Government Printing Office, 1939).

Grassett, Kurtis J. ["Paving Primer"], Town of Hancock, N.H., Department of Public Works, [2011], http://www.hancocknh.org/DPW/DPW-Road-Plan -2011.htm.

Petroski, Henry. *Paperboy: Confessions of a Future Engineer* (New York: Alfred A. Knopf, 2002).

Seely, Bruce E. *Building the American Highway System: Engineers as Policy Makers* (Philadelphia: Temple University Press, 1987).

12. THE PASSING THERE

Brandow, Michael. *New York's Poop Scoop Law: Dogs, the Dirt, and Due Process* (West Lafayette, Ind.: Purdue University Press, 2008).

DeJean, Joan. *How Paris Became Paris: The Invention of the Modern City* (New York: Bloomsbury, 2014).

Eno, William Phelps, and C. J. Tilden. *Sidewalks* (New Haven, Conn.: Eno Foundation for Highway Traffic Regulation, [1935]).

Holley, I. B., Jr. *The Highway Revolution, 1895–1925: How the United States Got Out of the Mud* (Durham, N.C.: Carolina Academic Press, 2008).

Judson, William Pierson. *City Roads and Pavements Suited to Cities of Moderate Size,* second edition and fourth edition, revised (New York: Engineering News Publishing, 1902 and 1909).

Lee, Jennifer S. "When Horses Posed a Public Health Hazard," *New York Times*, June 9, 2008, http://cityroom.blogs.nytimes.com/2008/06/09/when-horses-posed-a-public-health-hazard/?_r=0.

McCallum, Kevin. "Engineering Miscalculation Shuts Down Santa Rosa Project," (Santa Rosa, Calif.) *Press Democrat,* September 8, 2014, http://enr.construction.com/yb/enr/article.aspx?story_id=id:r-vhrIKAtGupa MIH4qC7MEVPcJBqeY4y1gnIZjdO8saJoo9_elZMHojLIazbnNL1&elq =70224609b9de4113918dea0e04adbe65&elqCampaignId=631.

McShane, Clay, and Joel A. Tarr. "The Centrality of the Horse in the Nineteenth-Century American City," in Raymond A. Mohl, ed., *The Making of Urban America* (Wilmington, Del.: Scholarly Resources, 1997), 105–130.

Mom, Gijs P. A., and David A. Kirsch. "Technologies in Tension: Horses, Electric Trucks, and the Motorization of American Cities, 1900–1925," *Technology and Culture* 42, no. 3 (July 2001): 489–518.

Montgomery, David. "After Gobbling Up Corvettes at Museum, Sinkhole Becomes a Star," *New York Times*, June 27, 2014, A22.

Williams, Lena. "With More Dog Waste on Streets, Crackdown Vowed," *New York Times*, April 19, 1997.

Woodruff, Cathy. "Granite Curbing Knifes Tires," *Albany (N.Y.) Times Union*, June 30, 2011, http://www.timesunion.com/local/article/Granite-curbing-knifes-tires-1446658.php.

13. HAD WORN THEM REALLY

Diana, Chelsea. "USM Heeds Pleas for Pine Siding, Not Vinyl, on 193-Year-Old Art Gallery," *Portland (Maine) Press Herald*, August 4, 2014, http://www.pressherald.com/2014/08/04/following-critics-meltdown-usm-heeds-pleas-to-use-pine-siding-not-vinyl-on-193-year-old-art-gallery/.

Kaysen, Ronda. "New, but Far From Perfect," *New York Times*, March 8, 2015, Business Section, 8.

Keenan, Sandy. "Treating His House Like a Museum," *New York Times*, August 7, 2014, D1, D7.

Krugman, Paul. "These Ages of Shoddy," *The Conscience of a Liberal* (blog), *New York Times,* June 27, 2014, http://krugman.blogs.nytimes.com/2014/06/27/these-ages-of-shoddy/?_php=true&_type=blogs&src =twr&_r=0.

Merkel, Jayne. "Try a D.I.Y. Fix for Public Housing," *New York Times*, September 16, 2014, A23.

Morgan, Curtis. "Impact of Hurricane Andrew: Better Homes," *Miami Herald*, August 19, 2014, http://www.miamiherald.com/news/special-reports/hurricane-andrew/article1940341.html.

Petroski, Henry. *The House with Sixteen Handmade Doors: A Tale of Architectural Choice and Craftsmanship* (New York: W. W. Norton, 2014).

———. "They Don't Make 'Em Like They Used To," *New York Times*, June 27, 2014, A29.

Rose, Ernestine Bradford. *The Circle: "The Center of Our Universe"* (Indianapolis: Indiana Historical Society, 1957).

Tsikoudakis, Mike. "Hurricane Andrew Prompted Better Building Code Requirements," *Business Insurance*, August 19, 2012, http://www.business-insurance.com/article/20120819/NEWS06/308199985.

Tuchman, Barbara W. "The Decline in Quality," *New York Times Magazine*, November 2, 1980, 38.

Wikipedia. "List of Frank Lloyd Wright Works," accessed November 13, 2014, http://en.wikipedia.org/wiki/List_of_Frank_Lloyd_Wright_works.

14. ABOUT THE SAME

Brown, John K. "Not the Eads Bridge: An Exploration of Counterfactual History of Technology," *Technology and Culture* 55, no. 3 (July 2014): 521–59.

Egan, Paul. "Michigan's $50 Fee for Super-Heavy Truck Loads Doesn't Go Far to Cover Stress on Roads," *Detroit Free Press*, August 5, 2013, http://archive.freep.com/article/20130804/NEWS06/110110010/Michigan-s-50-fee-super-heavy-truck-loads-doesn-t-go-far-cover-stress-roads.

Ferguson, Eugene S. "The American-ness of American Technology," *Technology and Culture* 20, no. 1 (January 1979): 3–24.

Gray, Kathleen. "Bid to Cut Truck Weights on Michigan Roads Fails," *Detroit Free Press*, December 2, 2014, http://www.freep.com/story/news/politics/2014/12/02/go-cutting-truck-weights/19787147/.

Petroski, Henry. *Engineers of Dreams: Great Bridge Builders and the Spanning of America* (New York: Alfred A. Knopf, 1995).

———. *Success Through Failure: The Paradox of Design* (Princeton, N.J.: Princeton University Press, 2006).

——. *To Forgive Design: Understanding Failure* (Cambridge, Mass.: Harvard University Press, 2012).

Shadwell, Arthur. *Industrial Efficiency: A Comparative Study of Industrial Life in England, Germany and America*, two volumes (London: Longmans, Green, 1906).

15. LAY IN LEAVES

Arrowsic, Town of. *2013–2014 Preliminary Report of the Municipal Officers of the Town of Arrowsic, Maine*, 2014.

Cho, Aileen, and Jeff Rubenstone. "Asphalt Prices Begin to Decline in Short Term," ENR.com, March 24, 2015, http://enr.construction.com/economics/quarterly_cost_reports/2015/0330-asphalt-prices-begin-to-decline-in-short-term.asp.

Grassett, Kurtis J. Letter re: Maintaining Our Road Network. Town of Hancock, N.H., Department of Public Works, n.d.

Kuennen, Tom. "Pavement Preservation: Techniques for Making Roads Last," *Asphalt*, Fall 2005, http://www.fhwa.dot.gov/pavement/preservation/ppc0605.cfm.

——. "Taming Disruptive Cracks to Preserve Pavement," *Better Roads*, August 2009, 16–22, http://www.expresswaysonline.com/pdf/BR%20RS%20AUG%2009%20CRACKING.pdf.

Petroski, Henry. *The House with Sixteen Handmade Doors: A Tale of Architectural Choice and Craftsmanship* (New York: W. W. Norton, 2014).

16. TRODDEN BLACK

Anderson, Jack. "The Great Highway Robbery," *Parade*, February 4, 1962, 18–20.

Associated Press. "U.S. Charges World Trade Center Steel Erector with DBE Fraud," ENR.com, July 31, 2014, http://enr.construction.com/yb/enr/article.aspx?story_id=id:fAcbTflerXKUsZkrDz4x_Uwj76BpGagIoLucGtfq2qVrsZe3gG6sPzscUs_AIFcizk1VTnPRdXf42Akh5zPHZQ**.

Bel Geddes, Norman. *Magic Motorways* (New York: Random House, 1940).

Federal Highway Administration. "Highway History: The Golden Fleece Why Was the $27 Billion Estimate So Wrong?," http://www.fhwa.dot.gov/infrastructure/50estimate.cfm.

Feuer, Alan. "Agency with a History of Graft and Corruption," *New York Times*, April 23, 2008. http://www.nytimes.com/2008/04/23/nyregion/23 buildings.html?_r=0.

Florman, Samuel C. *Good Guys, Wiseguys, and Putting Up Buildings: A Life in Construction* (New York: St. Martin's Press, 2012).

Gordon, Michael, and Elizabeth DePompei. "Paving Company Executive Pleads Guilty in Massive Fraud Case," *Charlotte (N.C.) Observer*, July 22, 2014, http://www.charlotte.observer.com/2014/07/22/5059471/paving -company-executive-pleads.html#.

Heiss, Laurie, and Jill Smyth. *The Merritt Parkway: The Road that Shaped a Region* (Charleston, S.C.: The History Press, 2014).

Swift, Earl. *The Big Roads: The Untold Story of the Engineers, Visionaries, and Trailblazers Who Created the American Superhighways* (Boston: Houghton Mifflin Harcourt, 2011).

Weingroff, Richard F. "The Battle of Its Life," *Public Roads*, May/June 2006, http://www.fhwa.dot.gov/publications/publicroads/06may/05.cfm.

17. FOR ANOTHER DAY

American Association of State Highway and Transportation Officials. "Matrix of Illustrative Surface Transportation Revenue Options," 2014, http://www .tripnet.org/docs/AASHTO_Matrix_of_Revenue_Options_2014-03-25.pdf.

Barro, Josh. "'Pension Smoothing': A Gimmick Both Parties Love," *New York Times*, July 31, 2014, A3.

Federal Highway Administration. *America's Highways, 1776–1976: A History of the Federal-Aid Programs* (Washington, D.C.: U.S. Department of Transportation, [1976]).

Kaszynski, William. *The American Highway: The History and Culture of Roads in the United States* (Jefferson, N.C.: McFarland, 2000).

Kile, Joseph. "The Highway Trust Fund and Paying for Highways," Testimony before the U.S. Senate Committee on Finance (Washington, D.C.: Congressional Budget Office, May 17, 2011).

———. "The Status of the Highway Trust Fund and Options for Financing Highway Spending," Testimony before the U.S. Senate Committee on Finance (Washington, D.C.: Congressional Budget Office, May 6, 2014).

Lee, Don. "Shoddy U.S. Roads and Bridges Take a Toll on the Economy," *Los Angeles Times*, August 14, 2014, http://www.latimes.com/national/la-na -roads-economy-20140815-story.html.

Pew Charitable Trusts. "Intergovernmental Challenges in Surface Transportation Funding," first report in *Fiscal Federalism in Action Series*, September 2014.

Povich, Elaine S. "Shrinking Revenue Spurs Gas Tax Alternatives," Pew Charitable Trusts, Stateline, August 14, 2014, http://www.pewtrusts.org /en/research-and-analysis/blogs/stateline/2014/08/14/shrinking-revenue-spurs-gas-tax-alternatives.

Scott, Robert III. "Fill 'Er Up: A Study of Statewide Self-Service Gasoline Station Bans," *Challenge* 50, no. 5 (September 2007): 103–114.

Swift, Earl. *The Big Roads: The Untold Story of the Engineers, Visionaries, and Trailblazers Who Created the American Superhighways* (Boston: Houghton Mifflin Harcourt, 2011).

Weingroff, Richard F. "Busting the Trust: Unraveling the Highway Trust Fund, 1968–1978" (Washington, D.C.: Federal Highway Administration, June 2013).

18. HOW WAY LEADS ON TO WAY

American Society of Civil Engineers. *Failure to Act Economic Studies* (Reston, Va.: ASCE, 2011–2013). http://www.asce.org/failuretoact/.

———. *Failure to Act: The Impact of Current Infrastructure Investment on America's Economic Future* (Reston, Va.: ASCE, 2013).

Bellisle, Martha. "Gov. Inslee Announces 12-Year, $12 Billion Plan for Transportation Infrastructure," *Olympia (Wash.) Olympian*, December 16, 2014, http://www.foxbusiness.com/markets/2014/12/16/gov-inslee-announces -12-year-12-billion-plan-for-transportation-infrastructure/.

Krugman, Paul. "Build We Won't," *New York Times*, July 4, 2014, A19.

Laing, Keith. "Poll: 67 Percent Oppose Gas Tax Hike," *Hill*, December 19, 2014, http://thehill.com/policy/transportation/227682-poll-67-percent-oppose-gas -tax-hike.

Roberts, Alasdair. *America's First Great Depression: Economic Crisis and Political Disorder after the Panic of 1837* (Ithaca, N.Y.: Cornell University Press, 2012), 51–52, 63–66.

Young, Virginia. "Missouri Transportation Tax Proposal Soundly Defeated," *St. Louis Post-Dispatch*, August 6, 2014, http://www.stltoday.com/news /local/govt-and-politics/missouri-transportation-tax-proposal-soundly -defeated/article_e3e0b7e6-4f55-5697-b759-d289d298d911.html.

19. EVER COME BACK

I am grateful to Don deKoven of Poughkeepsie, New York, for suggesting that I write about the local railroad bridge and its transformation, and for his helpful comments on the manuscript of my original essay on which this chapter was based. For background on the bridge's design and construction, I relied heavily on the Historic Civil Engineering Landmark nomination package prepared by Frank Griggs and submitted by the Mid-Hudson Branch of the Mohawk Hudson Section of the American Society of Civil Engineers to the society's History and Heritage Committee, which I chaired at the time.

Cardwell, Diane. "For High Line Visitors, Park Is a Railway out of Manhattan," *New York Times*, July 22, 2009, A17, A20.

Clarke, Thomas C. "The Hudson River Bridge at Poughkeepsie, N.Y.," *Scientific American Supplement*, no. 646, May 19, 1888, 10311–10313.

Cooper, C. Michael. "Walkway over the Hudson: Creating the World's Longest Pedestrian Bridge (Part 1)," *Rochester Engineer*, April 2009, 8–9.

Griggs, Francis E., Jr. "Thomas W. H. Moseley & His Bridges," *Civil Engineering Practice*, Fall/Winter 1997, 19–38.

———. "The Poughkeepsie Bridge: Its Birth, Abandonment and Rebirth," *Journal of Bridge Engineering* 14, no. 6 (November/December 2009): 518–28.

Hall, Christopher. "Bridge Makeover Collects Design Ideas," *Louisville (Ky.) Courier-Journal*, June 6, 2009, http://archive.courier-journal.com/article/20090606/NEWS02/90606006/Bridge-makeover-collects-design-ideas.

Jackson, Donald C. *Great American Bridges and Dams* (Washington, D.C.: Preservation Press, 1988).

Mabee, Carleton. *Bridging the Hudson: The Poughkeepsie Railroad Bridge and Its Connecting Rail Lines: A Many-Faceted History* (Fleischmanns, N.Y.: Purple Mountain Press, 2001).

Melewski, Peter, John Brizzell, Mike Cooper, et al. "Walking the Line," *Civil Engineering*, December 2009, 64–71.

Mort, Mike. *A Bridge Worth Saving: A Community Guide to Historic Bridge Preservation* (East Lansing: Michigan State University Press, 2008).

O'Rourke, John F. "The Construction of the Poughkeepsie Bridge," *Transactions of the American Society of Civil Engineers* 18 (1888): 199–216.

Pogrebin, Robin. "Renovated High Line Now Open for Strolling," *New York Times*, June 9, 2009, C3.

Wolf, Donald E. *Crossing the Hudson: Historic Bridges and Tunnels of the River* (New Brunswick, N.J.: Rutgers University Press, 2010).

20. TELLING THIS WITH A SIGH

Ambassador Bridge. "North America's #1 International Border Crossing," http://www.ambassadorbridge.com.

Boselovic, Len. "The 'P3' Dilemma," four-part series, *Pittsburgh Post-Gazette*, August 11–14, 2014.

Coyne, Justine. "PennDOT Picks Team to Replace 558 Bridges Through Rapid Bridge Program," *Pittsburgh Business Times*, October 24, 2014, http://www.bizjournals.com/pittsburgh/news/2014/10/24/penndot-picks-team-to-replace-558-bridges-through.html?s=print.

Detroit International Bridge Company. "Gateway Completion Statement," news release, September 20, 2012, http://www.ambassadorbridge.com/!Downloads/Gateway%20Completion%20Statement.pdf.

Dresden, Matthew. "Must a Bridge Be Beautiful Too?" *Access*, no. 28 (Spring 2006), 10–17.

Gallagher, John. "CEO on New Bridge to Windsor: 'It Will Happen.'" *Detroit Free Press*, August 24, 2014, http://www.freep.com/article/20140824/BUSINESS06/308240055/Detroit-Windsor-bridge-Cautillo.

Mason, Philip P. *The Ambassador Bridge: A Monument to Progress* (Detroit: Wayne State University Press, 1987).

Morris, Eric. "How Privatization Became a Train Wreck," *Access*, no. 28 (Spring 2006), 18–25.

Nussbaum, Paul. "Chinese Investors Sign Up to Fund I-95-Pa. Turnpike Link," *Philadelphia Inquirer*, November 30, 2014, http://www.philly.com/philly/business/20141130_Chinese_investors_sign_up_to_fund_I-95-Pa__Turnpike_link.html.

Panel on Public-Private Partnerships. *Public Private Partnerships: Balancing the Needs of the Public and Private Sectors to Finance the Nation's Infrastructure* (U.S. House of Representatives, Committee on Transportation and Infrastructure, 2014), http://transportation.house.gov/uploadedfiles/p3_panel_report.pdf?utm_campaign=GR-20140919-TWiW%20Email&utm_medium=email&utm_source=Eloqua.

Roumeliotis, Greg, and Mike Stone. "Exclusive: Infrastructure Investors Line Up for Indiana Toll Road," Reuters, October 14, 2014, http://www.reuters.com /article/2014/10/14/us-indianatollroad-m-a-idUSKCN0I32T720141014.

Schmitz, Jon. "Partnership to Replace 558 Bridges in Pa., Dozens of Small Ones in Allegheny County," *Pittsburgh Post-Gazette*, October 24, 2014, http ://www.post-gazette.com/news/transportation/2014/10/24/Pennsylvania-Department -of-Transportation-selects-partnership-to-replace-558-bridges/stories /201410240190.

Turner, Mike. "Matthew Moroun Speaks Out Against a New Bridge," *Corp Magazine*, September 8, 2011, http://www.corpmagazine.com/features /cover-stories/matthew-moroun-speaks-out-against-a-new-bridge-leave -crossings-to-the-private-sector-son-of-ambassador-bridge-owner-urges.

21. AGES AND AGES HENCE

American Society of Civil Engineers. *Failure to Act: The Impact of Current Infrastructure Investment on America's Economic Future* (Reston, Va.: ASCE, 2013).

Bliss, Laura. "The New Alchemy: How Self-Healing Materials Could Change the World," *The Atlantic* CityLab, September 15, 2014. http://www.citylab .com/tech/2014/09/the-new-alchemy-how-self-healing-materials -could-change-the-world/380075/.

DeLuca, Nick. "The Latest Innovative App from New Urban Mechanics Is Pioneering Data Donorship," BostInno, May 8, 2014, http://bostinno .streetwise.co/2014/05/08/boston-mayors-office-of-new-urban-mechanics -streetbump-app-2/.

———. "Northeastern Engineers Have Found a Way to Render Boston Potholes Extinct," BostInno, August 4, 2014, http://bostinno.streetwise.co/2014/08 /04/northeastern-engineers-have-found-a-way-to-render-boston-potholes -extinct/.

Durbin, Dee-Ann. "Self-Driving Car Safely Makes Trip from San Francisco to New York," *Salt Lake Tribune*, April 2, 2015, http://www.sltrib.com/csp /mediapool/sites/sltrib/pages/printfriendly.csp?id=2360219.

Frost, Robert. "Stopping by Woods on a Snowy Evening." Poetry Foundation, http://www.poetryfoundation.org/poem/171621.

Gardner, Bill. "The End of Potholes? UK Scientists Invent 'Self-Healing Concrete'," *(U.K.) Telegraph*, December 2, 2014, http://www.telegraph.co

.uk/news/uknews/road-and-rail-transport/11268310/The-end-of-potholes
-UK-scientists-invent-self-healing-concrete.html.

Gibson, Tom. "Short Road to the Next Ride," *Mechanical Engineering*, February
2015, 40–45.

Kessler, Aaron M. "Cars Conversing with Cars," *New York Times*, August 21,
2014, B1, B6.

Miodownik, Mark. *Stuff Matters: Exploring the Marvelous Materials That
Shape Our Man-Made World* (Boston: Houghton Mifflin Harcourt, 2013).

New Urban Mechanics. http://www.newurbanmechanics.org/.

Ngowi, Rodrique. "App Detects Potholes, Alerts Boston City Officials," *USA
Today*, July 20, 2012, http://usatoday30.usatoday.com/tech/news/story
/2012-07-20/pothole-app/56367586/1.

Rose, David. *Enchanted Objects: Design, Human Desire, and the Internet of
Things* (New York: Scribner, 2014).

Urmson, Chris. "Progress in Self-Driving Vehicles," *Bridge*, Spring 2015, 3–8.

List of Illustrations, with Credits

Index

Italicized page numbers refer to illustrations and their captions.

Index

A Note on the Author

Henry Petroski is the Aleksandar S. Vesic Professor of Civil Engineering and a professor of history at Duke University. He is the author of eighteen previous books, including *The Pencil: A History of Design and Circumstance*; *To Engineer Is Human: The Role of Failure in Successful Design*; *Engineers of Dreams: Great Bridge Builders and the Spanning of America*; and *The Essential Engineer: Why Science Alone Will Not Solve Our Global Problems*. Petroski lectures around the world on the material in his books, which have been translated into more than a dozen languages. He lives in North Carolina and Maine.